Math Kangaroo USA
Problems and Solutions

Grades 3 & 4

Odd Years
1999–2023

Editor in Chief

Agata Gazal
Chief Editorial Officer for Math Kangaroo USA
Billings, MT

Reviewers and Contributors

Joanna Matthiesen
Chief Executive Officer for Math Kangaroo USA
Granger, IN

Izabela Szpiech
Chief Financial Officer for Math Kangaroo USA
Chicago, IL

Kasia Nalaskowska
Chief Information Officer for Math Kangaroo USA
Aurora, IL

Magdalena Teodorowicz
Chief Design Officer for Math Kangaroo USA
Cordova, TN

Professor Andrzej Zarach, Ph.D.
Math Content Reviewer
East Stroudsburg University, East Stroudsburg, PA

Book Design

Jossea K. Rilea
Designer at LX Design Lab
Saratoga Springs, NY

We would like to give special thanks to other countless people who contributed to the problems and solutions for this book since 1998. Primarily, a big thank you to the Math Kangaroo question writers from all over the world that are part of the AKSF organization (www.aksf.org). Math Kangaroo solution writers also include Math Kangaroo USA competition organizers and Math Kangaroo Alumni. We would also like to thank the hundreds of educators who gave us feedback on the questions and solutions and finally the hundreds of thousands of students that took the Kangaroo challenge over the last two decades. Thank you all for your help in developing this book.

Copyright © 2024 by Math Kangaroo USA, NFP, Inc.
www.mathkangaroo.org

For additional copies of this book, please contact the publisher:
Math Kangaroo USA
info@mathkangaroo.org

ISBN: 979-8-98-998831-0

Preface

Hop to the Top with This Official Math Kangaroo Book for Grades 3 and 4!

Our newest preparation book for the annual Math Kangaroo Competition promises to spark a child's interest in mathematics — it introduces motivated elementary school students to math riddles, puzzles, and creative logic questions in fun and enjoyable ways. Math Kangaroo books do not focus on teaching arithmetic; instead, they are powerful tools in enhancing students' abilities to analyze and solve complex problems while using logical reasoning.

Inside the book, 3rd and 4th grade students will find 312 entertaining problems presented during actual Math Kangaroo Competition odd years, spanning 1999-2023, for a total of 13 tests, and their solutions. Each test consists of 24 questions divided into color-coded easy, medium, and difficult categories.

The Math Kangaroo question creation and selection process is intrinsically diverse, coming from the minds of top-university professors from over 100 countries worldwide. The best of the best questions, derived from group scrutiny, are selected each year at the Kangourou sans Frontières meeting. Out of thousands of submissions, only the best-crafted, most engaging, and age-appropriate questions are selected. After the selection process, solutions to the questions are written by a devoted team of Math Kangaroo USA faculty members, professors, and alumni.

This easy-to-use resource book will help students practice their math skills and also use math and logic as a tool for understanding the world around them.

We hope this book will be cherished by students who love mathematics, parents who like to study math with their children at home, and educators passionate about teaching unconventional and challenging math.

Enjoy this book and please let us know how you like it at www.mathkangaroo.org.

Joanna Matthiesen
President and CEO, Math Kangaroo USA

COLOR KEY

Each test has 24 questions with 3 levels of difficulty

GREEN	YELLOW	RED
Easy	Medium	Difficult
Q 1-8	Q 9-16	Q 17-24
3 Points	4 Points	5 Points

TABLE OF CONTENTS

Part I PROBLEMS..7
1999..9
2001..15
2003..21
2005..27
2007..33
2009..39
2011..45
2013..51
2015..57
2017..63
2019..69
2021..75
2023..81

Part II SOLUTIONS..87
1999..89
2001..95
2003..101
2005..107
2007..113
2009..119
2011..125
2013..131
2015..137
2017..143
2019..149
2021..155
2023..161

Part III ANSWER KEY..167

Part I
Problems

1999

1999

3 Points Each

1. Beata has two dolls, three apples, one chocolate bar, two oranges, five peaches, and one bike. How many pieces of fruit does Beata have?

   (A) 3
   (B) 5
   (C) 10
   (D) 18
   (E) 21

2. What number is in the part that is common to four circles?

   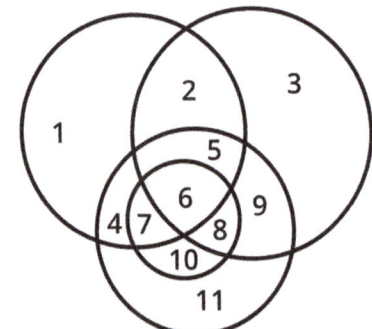

   (A) 5
   (B) 9
   (C) 7
   (D) 4
   (E) 6

3. In how many places do we need to break a wooden stick in order to get 5 pieces?

   (A) 3
   (B) 4
   (C) 5
   (D) 6
   (E) It depends on how long the stick is.

4. Karl is now 10 years old, and Alice is 3 years old. How many years from now will Karl be twice as old as Alice?

   (A) 5
   (B) 10
   (C) 4
   (D) 1
   (E) 3

5. Anna and her sister Barbara go to the same school, but take two different ways to get to school. Whose way is shorter?

   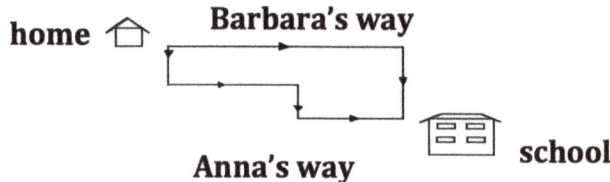

   (A) Anna's way
   (B) Barbara's way
   (C) It depends on the distance to the school.
   (D) Both ways have the same length.
   (E) It is impossible to determine.

6. Our class has 30 students. The number of boys is four times the number of girls. How many girls are there in our class?

   (A) 24
   (B) 16
   (C) 12
   (D) 8
   (E) 6

7. How much does the orange weigh?

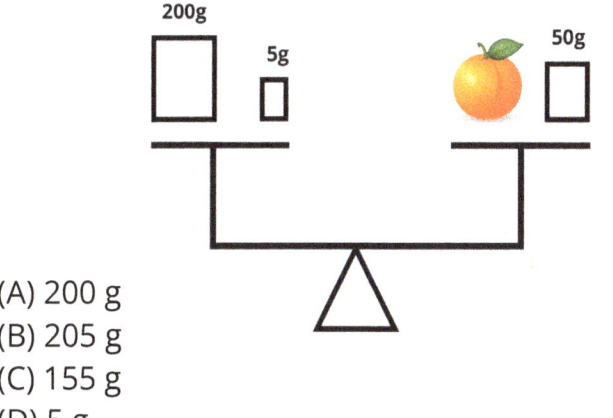

(A) 200 g
(B) 205 g
(C) 155 g
(D) 5 g
(E) It cannot be determined.

8. The cat says, "The length of my tail is 12 cm and half the length of my tail." How long is the cat's tail?

(A) 18 cm
(B) 24 cm
(C) 12 cm
(D) 9 cm
(E) 6 cm

4 Points Each

9. My mom's birthday is on a Sunday, and my dad's birthday is 55 days later. On what day of the week will my dad's birthday be?

(A) Sunday
(B) Monday
(C) Tuesday
(D) Thursday
(E) Saturday

10. Two basketball teams are playing against each other. The first team to win four games wins the tournament. There are no ties. What is the greatest number of games that could take place after which the winner will be known for sure?

(A) 8
(B) 7
(C) 6
(D) 5
(E) 4

11. Instead of adding 27 to a certain number, John subtracted 27 from that number. What is the difference between John's result and the result he should have gotten?

(A) 27
(B) 0
(C) 54
(D) 100
(E) 3

12. A goldsmith decided to cut a golden cube with an edge of 4 cm into small cubes each with an edge of 1 cm. How many small cubes will he have?

(A) 64
(B) 48
(C) 32
(D) 16
(E) 12

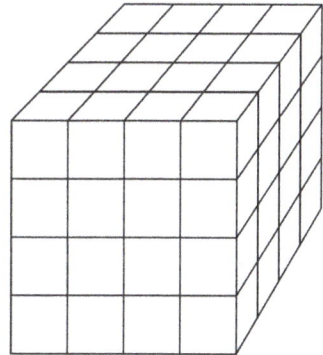

13 A pail filled with milk to the top weighs 25 kilograms, and a pail filled half-way weighs 13 kilograms. How much does an empty pail weigh?

(A) 2 kilograms
(B) ½ kilogram
(C) 1 ½ kilograms
(D) 1 kilogram
(E) 2 ½ kilograms

14 In Grandma's pantry, there is a jar with 650 g of jam. Each day her grandson Tom eats 5 teaspoons of jam from the jar. Each teaspoon holds 6 g of jam. How much jam will there be left in the jar after 20 days?

(A) 50 g
(B) 530 g
(C) 550 g
(D) 1250 g
(E) The jar will be empty.

15 Each of the kangaroo's eleven children has eleven children, and each of them also has eleven children. How many great-grandchildren does the kangaroo have?

(A) 111
(B) 121
(C) 11211
(D) 1331
(E) 12321

16 What is the least possible number of children in the Kowalski family if each of the children has at least one brother and at least one sister?

(A) 1
(B) 2
(C) 3
(D) 4
(E) 5

5 Points Each

17 Peter opened a book and found that the sum of the page number on the left and the page number on the right is equal to 21. What is the product of the two page numbers?

(A) 121
(B) 100
(C) 420
(D) 110
(E) 426

18 Father Virgil is taking care of 143 children. Each day, each child gets half a liter of milk with breakfast. The milk from one cow is enough for 40 children. What is the least number of cows that Father Virgil needs to have?

(A) 2
(B) 3
(C) 4
(D) 5
(E) 6

19 A kangaroo wants to make a rectangular bedspread 1.5 m long and 1 m wide using square scraps which measure 10 cm × 10 cm. At every point where four squares meet she wants to place a fancy button. How many buttons will she need?

(A) 150
(B) 104
(C) 126
(D) 140
(E) 135

20 Pinocchio's wooden nose is 3 cm long. Whenever Pinocchio lies, the length of his nose doubles. How long will his nose be after he tells 6 lies?

(A) 192 cm
(B) 67 cm
(C) 96 cm
(D) 18 cm
(E) 384 cm

21 In the yard there is an equal number of pigs, ducks, and chickens. Together, they have 144 legs. How many ducks are there in the yard?

(A) 18
(B) 21
(C) 35
(D) 42
(E) 43

22 One number was chosen from the numbers 51, 52, 53, 54, and 55, and the digit 0 was placed between the digits of that number. What is the difference between the new number and the number which was chosen?

(A) 500
(B) 50
(C) 550
(D) 450
(E) The difference depends on which number was chosen.

23 If Grandma gave each of her grandchildren 10 pieces of candy, there would not be any candy left over for one of the grandchildren. If she gave each one of them 8 pieces of candy, she would have 6 pieces of candy left. How many grandchildren does she have?

(A) 4
(B) 6
(C) 8
(D) 10
(E) 12

24 The figure shown rotates clockwise and makes one full rotation in one hour. Its position at 12:00 p.m. is shown in the picture to the right. What will it look like at 2:15 p.m.?

(A)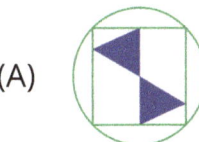

(B)

(C)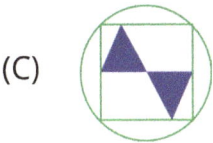

(D)

(E)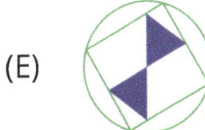

2001

2001

3 Points Each

1 Julia has four candles. Each of her candles burns itself out in three hours. Julia lit two candles and placed them next to an open window. 30 minutes later the wind blew out one candle, and an hour later the wind blew out the second candle. Then Julia closed the window and lit all four candles. How long after this moment will the last candle burn out?

(A) 1 hr 30 min
(B) 2 hr
(C) 3 hr
(D) 7 hr 30 min
(E) 8 hr

2 Joseph had 7 sticks. He broke one of them into two pieces. How many sticks does Joseph have now?

(A) 5
(B) 6
(C) 7
(D) 8
(E) 9

3 Kuba bought a chocolate heart for his mother (see the picture). Each chocolate square weighs 10 grams. What is the weight of the whole heart?

(A) 340 grams
(B) 360 grams
(C) 380 grams
(D) 400 grams
(E) 420 grams

4 There are 12 pairs of shoes on each of 10 shelves in the animal shoe store. 5 centipedes are the first customers. Three of them buy 30 pairs of shoes each, and two of them buy 5 pairs of shoes each. How many pairs of shoes are left on the shelves?

(A) 10
(B) 15
(C) 20
(D) 25
(E) 30

5 For five days, Kaya was helping her mother pick berries. On the first day, she ate most of her berries and gave her mother only one cup of berries. She decided that each day she would give her mother twice as much berries as the day before. How many cups of berries did Kaya give her mother over five days?

(A) 5
(B) 31
(C) 21
(D) 11
(E) 16

6 Which of the following expressions is correct?

(A) 12 ÷ (4 + 8) = 11
(B) 8 × 2 + 3 = 40
(C) 2 × 3 + 4 × 5 = 50
(D) (10 + 8) ÷ 2 = 14
(E) 18 − 6 ÷ 3 = 16

7. There are 19 girls and 12 boys in the school yard. At least how many students need to join them so that six groups with the same number of students can be formed?

(A) 1
(B) 2
(C) 3
(D) 4
(E) 5

8. Four sticks, each 14 cm long, were placed in the way shown in the picture, for a total length of 80 cm. The distances between the sticks are equal. How long is each of these distances?

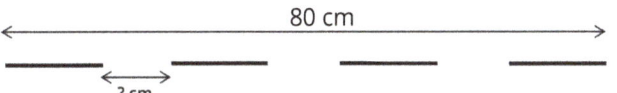

(A) 1 cm
(B) 2 cm
(C) 3 cm
(D) 5 cm
(E) 8 cm

4 Points Each

9. Bobby was born on Abby's third birthday. How many years later will Abby be twice as old as Bobby?

(A) 1
(B) 2
(C) 3
(D) 4
(E) 5

10. The lowest point of the Sniezna Cave in the Tatra Mountains is located 221 m below the cave entrance, and the highest point of the Sniezna Cave is located 419 m above the cave entrance. What is the depth (the distance between the lowest and highest points) of this cave?

(A) 198 m
(B) 221 m
(C) 419 m
(D) 640 m
(E) 650 m

11. I divided 20 pieces of candy among several children. Each child received at least one piece of candy, and everyone received a different number of pieces of candy. What is the greatest possible number of children who received candy?

(A) 20
(B) 10
(C) 8
(D) 6
(E) 5

12. Which figure is different from all the others?

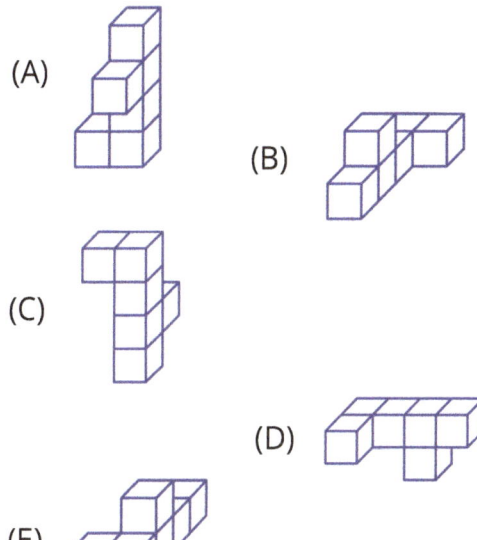

13 Ella and Bella got on a super-train. Ella took a seat in the seventeenth car counting from the front of the train, and Bella was seated in the thirty-fourth car counting from the end. The girls were sitting in the same car. How many cars did the super-train have?

(A) 48
(B) 49
(C) 50
(D) 51
(E) 52

14 Adam and Bart are picking up chestnuts. At a certain moment they have the same number of chestnuts, and then Adam gives Bart half of all his chestnuts. How many times is the number of chestnuts Bart has greater than the number of chestnuts Adam has now?

(A) 2 times
(B) 3 times
(C) 4 times
(D) 5 times
(E) It depends on the number of chestnuts they had at the beginning.

15 There are squares and triangles on the table. They have 17 vertices altogether. How many triangles are there on the table?

(A) 1
(B) 2
(C) 3
(D) 4
(E) 5

16 What is the least number of matches that must be added to the picture in order to get exactly 11 squares?

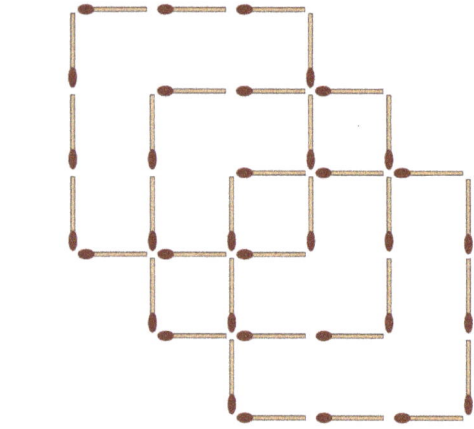

(A) 2
(B) 3
(C) 4
(D) 5
(E) 6

5 Points Each

17 In a certain picture you can see numbers

with their mirror reflections.

What is the next picture in the sequence of reflections?

(A)
(B)
(C)
(D)
(E)

18 Which of the five napkins below comes from this paper cut-out?

(A)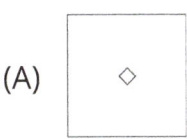

(B)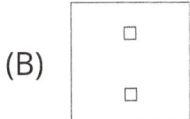

(C)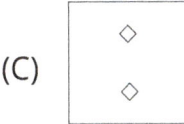

(D)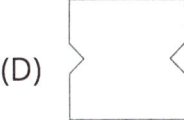

(E)

19 12 boys and 8 girls were members of the Math Club. Every week new members were admitted: 1 boy and 2 girls. How many members will this club have when the number of boys equals the number of girls?

(A) 20
(B) 24
(C) 28
(D) 32
(E) 36

20 How many three-digit numbers are there such that the sum of their digits equals 4?

(A) 10
(B) 9
(C) 8
(D) 7
(E) 6

21 Five girls made a square with their beach towels (see the picture). The towels of Anita and Beth are squares with a perimeter of 720 cm each. The towels of Cindy, Debbie, and Eva are identical rectangles. What is the perimeter of one of these rectangles?

(A) 600 cm
(B) 560 cm
(C) 440 cm
(D) 360 cm
(E) 300 cm

22 Todd has as much money as Will and Kevin have together. Will has 10 dollars less than Kevin. The three boys have 40 dollars altogether. How much money does Will have?

(A) $4
(B) $5
(C) $10
(D) $15
(E) $20

23 I have 11 candy bars in each of three baskets. I take out one candy bar from each basket in the following order: from the left, from the middle, from the right, from the middle, from the left, from the middle, from the right, and so on. What is the largest number of candy bars left in one of the baskets when the middle basket is empty?

(A) 1
(B) 2
(C) 5
(D) 6
(E) 11

24 The sum of the dots on the opposite sides of a die is seven. We move a die on a grid as the picture shows. The die is rotated once for each square as it is moved as shown by the arrows. How many dots are on the top of the die when it is located on the square marked with ✱?

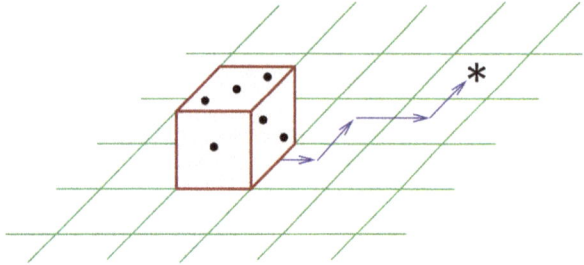

(A) 5
(B) 4
(C) 3
(D) 1
(E) other number

3 Points Each

1 The picture shows the letter U drawn on grid paper. How many squares does the letter U cover?

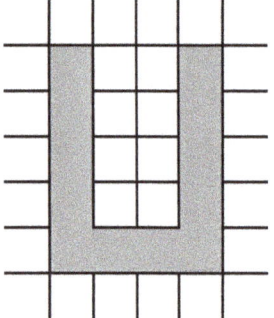

(A) 10
(B) 8
(C) 11
(D) 13
(E) 12

2 What is the result of
$0 + 1 + 2 + 3 + 4 - 3 - 2 - 1 - 0$?

(A) 0
(B) 2
(C) 4
(D) 10
(E) 16

3 The first car of the train, right behind the engine, contains 10 boxes. In each of the following cars there are twice as many boxes as in the car in front of it. How many boxes are there in the fifth car?

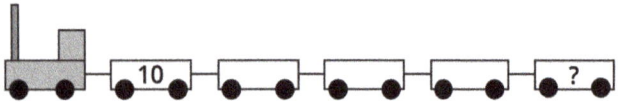

(A) 100
(B) 120
(C) 140
(D) 160
(E) 180

4 Sophia is drawing kangaroos. The first one is blue, the next one green, the one after it red, the fourth one yellow, and then again blue, green, red, yellow, and so on, in the same order. What color will the seventeenth kangaroo be?

(A) blue
(B) green
(C) red
(D) black
(E) yellow

5 In the teachers' lounge there are 6 tables with 4 chairs each, 4 tables with 2 chairs each, and 3 tables with 6 chairs each. How many chairs are there in the lounge?

(A) 40
(B) 25
(C) 50
(D) 36
(E) 44

6 In one of these pictures, there are three times as many hearts as other shapes. Which picture is it?

(A)

(B)

(C)

(D)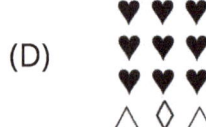

(E)

7. A rectangle with dimensions 7 × 4 was outlined on grid paper. How many squares will a diagonal of this rectangle intersect?

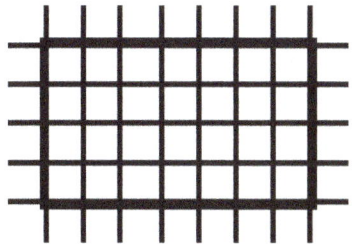

(A) 8
(B) 9
(C) 10
(D) 11
(E) 12

8. The figure presented in the picture, which is made out of a certain number of identical cubes, weighs 189 grams. How much does one cube weigh?

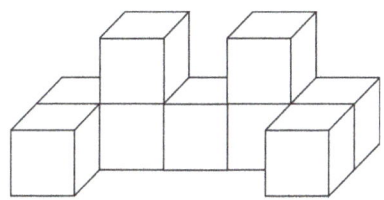

(A) 29 grams
(B) 25 grams
(C) 21 grams
(D) 19 grams
(E) 17 grams

4 Points Each

9. Philip wrote down consecutive natural numbers starting with 3 until he had written 35 digits. What was the greatest number that Philip wrote down?

(A) 12
(B) 22
(C) 23
(D) 28
(E) 35

10. Anna fell asleep at 9:30 p.m. and woke up at 6:45 a.m. the next day. Her little brother Peter slept 1 hour and 50 minutes longer. How long did Peter sleep?

(A) 30 hr 5 min
(B) 11 hr 35 min
(C) 11 hr 5 min
(D) 9 hr 5 min
(E) 8 hr 35 min

11. A pattern, the beginning and the end of which is shown in the picture, is made up of alternating black and white bars. There are 17 bars altogether. The bars on both ends are black. How many white bars are there in the pattern?

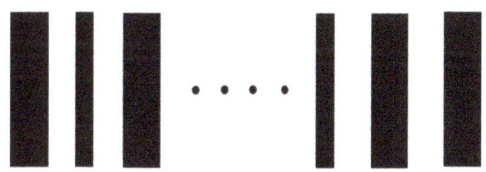

(A) 9
(B) 16
(C) 7
(D) 8
(E) 15

12. Jumping Kangaroo is practicing for the animal Olympics. His longest jump during the training was 55 dm 50 mm long, but in the finals of the Olympics he won with a jump that was 123 cm longer. How long was Jumping Kangaroo's longest jump during the Olympics?

(Remember that 1 m = 10 dm, 1 dm = 10 cm, and 1 cm = 10 mm)

(A) 6 m 78 cm
(B) 5 m 73 cm
(C) 5 m 55 cm
(D) 11 m 28 cm
(E) 7 m 23 cm

13 Paul chose a certain number, subtracted 203 from it, and then added 2003 to the difference. His final result was 20003. What number did Paul choose at the beginning?

(A) 23
(B) 17797
(C) 18203
(D) 21803
(E) 22209

14 Barbara's electronic watch shows time in 24-hour fomat. She likes to add the digits showing the current time. For example, when the watch shows 21:17, she gets the number 11 as the result. What is the greatest sum she can get?

(Hint: In some countries and sometimes in the US, instead of saying "It is 1 p.m.," people may say, "It is 13:00." When it is 2 p.m., they may say, "It is 14:00," and when it is 12 a.m., they may say, "It is 24:00." In this problem, 21:17 means 9:17 p.m. Time expressed in this way is sometimes called "military time.")

(A) 24
(B) 36
(C) 19
(D) 25
(E) 23

15 Mark said to his friends, "If I had picked twice as many apples as I did, I would have 24 more apples than I have now." How many apples did Mark pick?

(A) 48
(B) 24
(C) 42
(D) 12
(E) 36

16 Points A, B, C, and D, all of which lie on a straight line, are marked in the picture below. The distance between points A and C is 10 m, between B and D is 15 m, and between A and D is 22 m. What is the distance between points B and C?

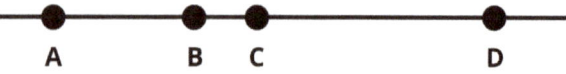

(A) 1 m
(B) 2 m
(C) 3 m
(D) 4 m
(E) 5 m

5 Points Each

17 There are 29 students in a certain class. 12 of the students have a sister and 18 of the students have a brother. In this class, only Tania, Barbara, and Anna do not have any siblings. How many students from this class have both a brother and a sister?

(A) none
(B) 1
(C) 3
(D) 4
(E) 6

18 Daniel has 11 pieces of paper. He cut some of them into three parts and now he has 29 pieces of paper. How many pieces of paper did he cut?

(A) 3
(B) 2
(C) 8
(D) 11
(E) 9

19) John bought 3 kinds of cookies: large, medium, and small. The large cookies cost 4 dollars each, the medium 2 dollars each, and the small 1 dollar each. Altogether, John bought 10 cookies and paid 16 dollars. How many large cookies did he buy?

(A) 5
(B) 4
(C) 3
(D) 2
(E) 1

20) Christopher built the rectangular prism shown in the picture using red and blue cubes of identical size. The outer walls of this prism are red but all the inner cubes are blue. How many blue cubes did Christopher use in this construction?

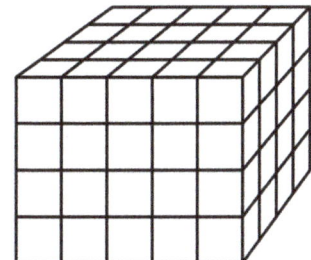

(A) 12
(B) 16
(C) 22
(D) 26
(E) 32

21) Jerry is planning to buy some basketballs. If he buys 5 basketballs, he will have 10 dollars left over, and if he buys 7 basketballs, he will have to borrow 22 dollars. How many dollars does one basketball cost?

(A) 11
(B) 16
(C) 22
(D) 26
(E) 32

22) Mark built a rectangular prism using 3 blocks, each of which is made out of 4 small cubes connected in various ways. The two colored blocks can be fully seen in the picture. Which is the third (white) block with only two visible sides?

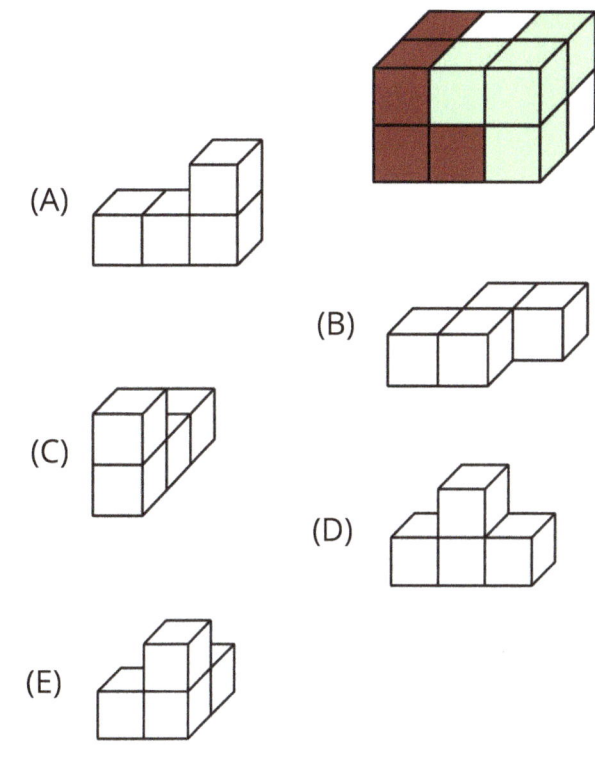

23) Two pieces that made up the shaded region were cut out from a square puzzle (see the picture). Which two of the pieces are they?

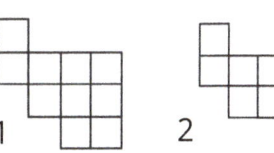 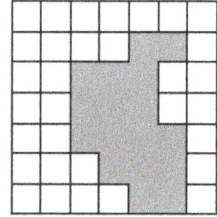

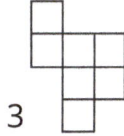 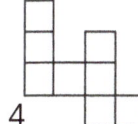

(A) 1 and 3
(B) 2 and 4
(C) 2 and 3
(D) 1 and 4
(E) 3 and 4

24 At the toy store, among other things, you can buy dogs, bears, and kangaroos. Three dogs and two bears together cost as much as four kangaroos. For the same amount of money, you can buy one dog and three bears. Which of the statements below is true?

(A) A dog is twice as expensive as a bear.
(B) A bear is twice as expensive as a dog.
(C) The prices of a dog and of a bear are identical.
(D) A bear is three times as expensive as a dog.
(E) A dog is three times as expensive as a bear.

2005

2005

3 Points Each

1 A butterfly sat down on a correctly solved problem. What number did it cover up?

$$2005 - 205 = 1300 +$$

(A) 250
(B) 400
(C) 500
(D) 910
(E) 1800

2 At noon, the minute hand of a clock is in the position shown in the picture on the right. What will the position of the minute hand be after 17 quarters of an hour pass?

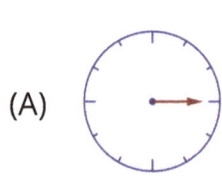

(A)

(B)

(C)

(D)

(E)

3 Joan bought some cookies, each of which costs 3 dollars. She gave the salesperson 10 dollars and received 1 dollar as change. How many cookies did Joan buy?

(A) 2
(B) 3
(C) 4
(D) 5
(E) 6

4 After the trainer's first whistle, the monkeys at the circus formed 4 rows. There were 4 monkeys in each row. After the second whistle, they rearranged themselves into 8 rows. How many monkeys were there in each row after the second whistle?

(A) 1
(B) 2
(C) 3
(D) 4
(E) 5

5 Eva lives with her parents, her brother, one dog, two cats, two parrots, and four fish. What is the total number of legs that they have altogether?

(A) 22
(B) 24
(C) 28
(D) 32
(E) 40

6. John has a chocolate bar made up of square pieces 1 cm × 1 cm in size. He has already eaten some of the corner pieces (see the picture). How many pieces does John have left?

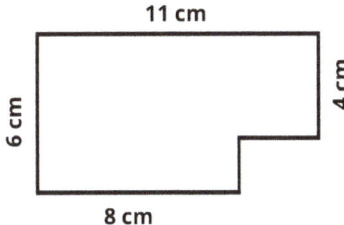

(A) 66
(B) 64
(C) 62
(D) 60
(E) 58

7. Two traffic signs mark the bridge in my village (see the picture below). These signs indicate the maximum vehicle width and the maximum vehicle weight allowed on the bridge. Which one of the following trucks is allowed to cross that bridge?

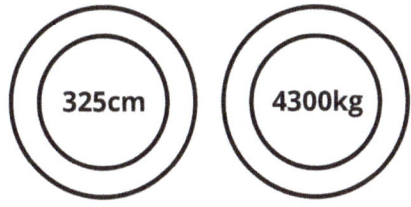

(A) a truck that is 315 cm wide and weighs 4400 kg
(B) a truck that is 330 cm wide and weighs 4250 kg
(C) a truck that is 325 cm wide and weighs 4400 kg
(D) a truck that is 330 cm wide and weighs 4200 kg
(E) a truck that is 325 cm wide and weighs 4250 kg

8. Each of seven boys paid the same amount of money for a trip. The total sum of what they paid is a three-digit number which can be written as 3☐0. What is the middle digit of this number?

(A) 3
(B) 4
(C) 5
(D) 6
(E) 7

4 Points Each

9. What is the smallest possible number of children in a family where each child has at least one brother and at least one sister?

(A) 2
(B) 3
(C) 4
(D) 5
(E) 6

10. From the five numbers below, I chose one number. The number is even and all of its digits are different. The hundreds digit is double the ones digit. The tens digit is greater than the thousands digit. Which number did I choose?

(A) 1246
(B) 3874
(C) 4683
(D) 4874
(E) 8462

11 A square piece of paper has been cut into three pieces. Two of them are shown in the picture below. Which of the pieces (A) to (E) is the third one?

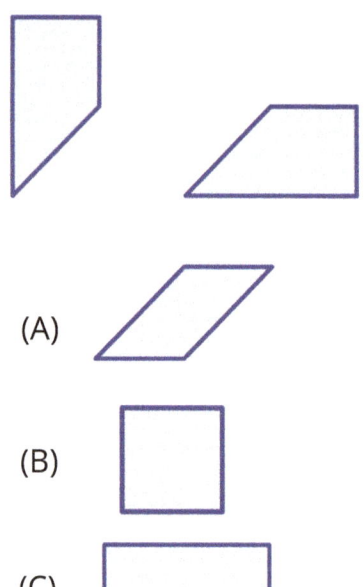

(A)

(B)

(C)

(D)

(E)

12 An elevator cannot carry more than 150 kg. Four friends weigh 60 kg, 80 kg, 80 kg, and 80 kg, respectively. What is the least number of trips necessary to carry the four friends to the highest floor?

(A) 1
(B) 2
(C) 3
(D) 4
(E) 7

13 Ala has 24 dollars and Barb has 66 dollars. Sophia has as much less than Barb as she has more than Ala. How many dollars does Sophia have?

(A) 33
(B) 35
(C) 42
(D) 45
(E) 48

14 There are eight kangaroos in the cells of the table (see the picture). What is the least number of kangaroos that need to be moved to the empty cells so that there would be exactly two kangaroos in any row and in any column of the table?

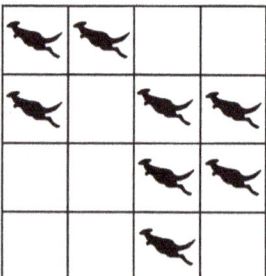

(A) 4
(B) 3
(C) 2
(D) 1
(E) 0

15 Greg has a sack with a hole. He needs to bring four full sacks of sand from the river to a house located at the other end of the village. Unfortunately, every time he goes through the village, half of the sand spills out of the sack through the hole. How many trips does Greg need to make from the river to the house in order to bring the required amount of sand?

(A) 4
(B) 5
(C) 6
(D) 7
(E) 8

16 During a Kangaroo camp, Adam solved five problems every day and Brad solved two problems daily. After how many days did Brad solve as many problems as Adam solved in 4 days?

(A) after 5 days
(B) after 7 days
(C) after 8 days
(D) after 10 days
(E) after 20 days

5 Points Each

17 There were 9 pieces of paper. Some of them were cut into three pieces. As a result, there are now 15 pieces of paper now. How many pieces of paper were cut?

(A) 2
(B) 3
(C) 4
(D) 5
(E) 6

18 Using 6 matches, only one rectangle with a perimeter of 6 matches can be made (see the picture). How many different rectangles with a perimeter of 14 matches can be made using 14 matches?

(A) 2
(B) 3
(C) 4
(D) 6
(E) 12

19 A picture frame was constructed using pieces of wood which all have the same width. What is the width of the frame if the inside perimeter of the frame is 8 decimeters (dm) less than its outside perimeter?

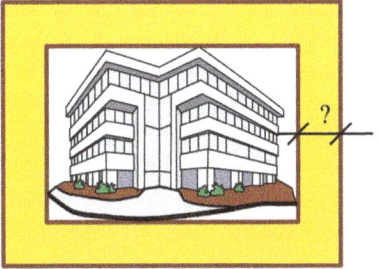

(A) 1 dm
(B) 2 dm
(C) 4 dm
(D) 8 dm
(E) It depends on the size of the picture.

20 In a certain trunk there are 5 chests, in each chest there are 3 boxes, and in each box there are 10 gold coins. The trunk, the chests, and the boxes are locked. At least how many locks need to be opened in order to take out 50 coins?

(A) 5
(B) 6
(C) 7
(D) 8
(E) 9

21 The figure shows a rectangular garden with dimensions of 16 m by 20 m. The gardener planted six identical flowerbeds (colored gray in the diagram). What is the perimeter of each of the flowerbeds?

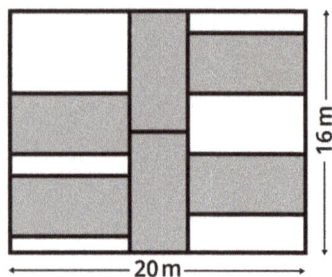

(A) 20 m
(B) 22 m
(C) 24 m
(D) 26 m
(E) 28 m

22 Mike chose a three-digit number and a two-digit number. The difference of these numbers is 989. What is their sum?

(A) 1001
(B) 1010
(C) 2005
(D) 1000
(E) 1009

23 Five cards are lying on the table in the order 5, 1, 4, 3, 2, as shown in the top row of the picture. They need to be placed in the order shown in the bottom row. In each move, any two cards may be switched. What is the least number of moves that need to be made?

| 5 | 1 | 4 | 3 | 2 |

| 1 | 2 | 3 | 4 | 5 |

(A) 2
(B) 3
(C) 4
(D) 5
(E) 6

24 Which of the cubes has the plan shown in the picture to the right?

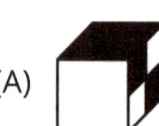 (A)

 (B)

 (C)

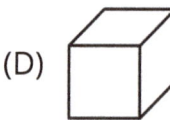 (D)

 (E)

2007

2007

3 Points Each

1 Not taking any steps backwards, Anna walked toward the car using a path shown in the picture, and picked up numbers she encountered along her way. Which set of the numbers below could she pick up?

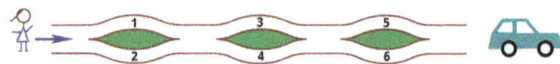

(A) 1, 2, 4
(B) 2, 3, 4
(C) 2, 3, 5
(D) 1, 5, 6
(E) 1, 2, 5

2 How many letters from the word KANGUR are repeated in the word PROBLEM?

(A) 1
(B) 2
(C) 3
(D) 4
(E) 5

3 Which of the patterns shown below consists of the largest number of squares?

(A)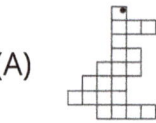

(B)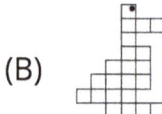

(C)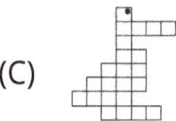

(D)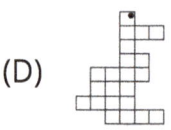

(E)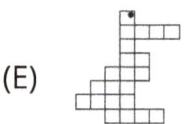

4 Helen has $5. She is going to buy 5 notebooks which cost 80 cents each and a certain number of pencils which cost 30 cents each. How many pencils at most can she buy?

(A) 5
(B) 4
(C) 3
(D) 2
(E) 1

5 There are 9 streetlights on one side of a path in the park. The distance between neighboring streetlights is 8 meters. Gregory went along this path from the first lantern to the last lantern. How many meters did he walk?

(A) 48
(B) 56
(C) 64
(D) 72
(E) 80

6. A 3-digit code is needed to open a safe. How many possible codes are there if it is known that only three numbers, 1, 3, and 5, are used in this code, and each of them is used only once?

(A) 2
(B) 3
(C) 4
(D) 5
(E) 6

7. 4 × 4 + 4 + 4 + 4 + 4 + 4 × 4 = ?

(A) 32
(B) 44
(C) 48
(D) 56
(E) 144

8. Which figure of those shown below can be placed next to the figure shown to the right in order to form a rectangle?

(A)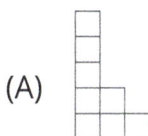

(B)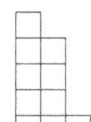

(C)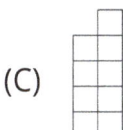

(D)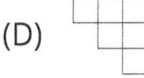

(E)

4 Points Each

9. What number needs to be written in the shaded cloud in order to get the number in the last cloud as the result of the operations shown in the picture?

? $\xrightarrow{-2}$ $\xrightarrow{\div 3}$ $\xrightarrow{+4}$ 5

(A) 1
(B) 3
(C) 5
(D) 7
(E) 9

10. The square shown in the picture must be filled in such a way that each of the digits 1, 2, and 3 appears in each row and in each column once and only once. If Harry started to fill in the square as shown, what number can he write in the square marked with the question mark?

1	?	
2	1	

(A) 1
(B) 2
(C) 3
(D) 1 or 2
(E) 1, 2, or 3

11 What is the smallest number greater than 2007 in which the sum of the digits is equal to the sum of the digits of 2007?

(A) 2016
(B) 2015
(C) 2009
(D) 1008
(E) 2070

12 Annette is putting identical cube blocks into a cube aquarium. She has already put in a certain number of blocks (look at the picture). How many such blocks does she still need to add to fill up the aquarium?

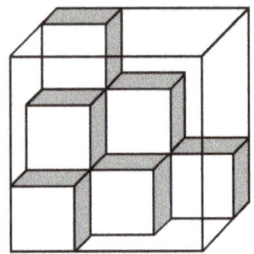

(A) 9
(B) 13
(C) 17
(D) 21
(E) 27

13 Peter, who is 1 year and 1 day older than Paul, was born on January 1st, 2002. When was Paul born?

(A) January 2nd, 2003
(B) January 2nd, 2001
(C) December 31st, 2000
(D) December 31st, 2002
(E) December 31st, 2003

14 A string has been cut into 400 pieces, each 15 cm long. How long was the string? (Note: 1 km = 1000 m, 1 m = 100 cm, 1 cm = 10 mm)

(A) 6 km
(B) 60 m
(C) 600 cm
(D) 6,000 mm
(E) 60,000 cm

15 David wrote a one-digit number, and next to it to the right he wrote another digit to form a two-digit number. Then he added 19 to this number and the sum was 72. What was the first digit he wrote?

(A) 2
(B) 5
(C) 6
(D) 7
(E) 9

16 An electronic watch indicates the time 02:07. After how much time will the same digits show up again for the first time, not necessarily in the same order?

(A) 4 hr 55 min
(B) 6 hr
(C) 10 hr 55 min
(D) 11 hr 13 min
(E) 24 hr

5 Points Each

17 A cube with an edge 3 cm long has been painted gray. Next, it has been cut into small cubes with an edge 1 cm long (see the picture). How many small cubes have exactly two gray sides?

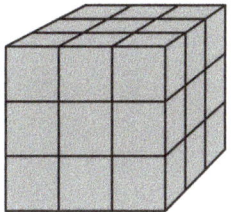

(A) 4
(B) 6
(C) 8
(D) 10
(E) 12

18 We call a number palindromic if it doesn't change after its digits are written in reverse order. Some examples are 1331 and 24642. A certain car's odometer shows 15951 kilometers. After how many more kilometers will a palindromic number show up on the odometer the very next time?

(A) after 100 km
(B) after 110 km
(C) after 710 km
(D) after 900 km
(E) after 1010 km

19 If you count the small white squares in the sequence of big squares shown in the pictures below, you will get the numbers listed.

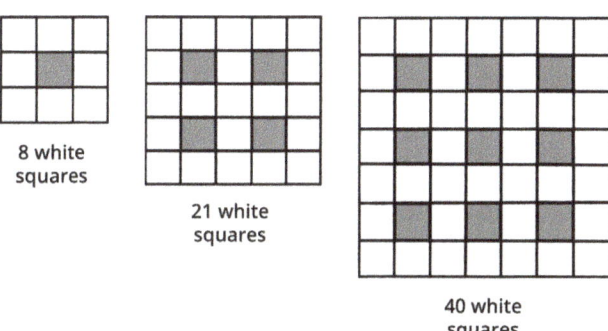

8 white squares

21 white squares

40 white squares

If we continue the pattern, how many white squares will there be in the next big square?

(A) 50
(B) 60
(C) 65
(D) 70
(E) 75

20 Adam, Bob, Celina, Daniel, and Eve formed a line to the cashier. Adam is standing farther from the cashier than Celina. Bob is standing closer to the cashier than Adam and right behind Daniel. Daniel is standing closer than Celina, but he isn't first in line. In what place, counting from the cashier, is Eve standing?

(A) 1st
(B) 2nd
(C) 3rd
(D) 4th
(E) 5th

Problems - 2007

21 In four corners of a rectangle with dimensions 15 cm × 9 cm, four squares each with a perimeter equal to 8 cm were cut out. What is the perimeter of the polygon created in this way?

(A) 48 cm
(B) 40 cm
(C) 32 cm
(D) 24 cm
(E) 16 cm

22 At a round table there are chairs placed with the same distance between them. They are numbered consecutively 1, 2, 3, … . Joe is sitting in the chair number 11, directly across from Chris, who is sitting in the chair number 4. How many chairs are there at the table?

(A) 13
(B) 14
(C) 16
(D) 17
(E) 22

23 How many digits have to be written in order to write down every number from 1 to 100 inclusive?

(A) 100
(B) 150
(C) 190
(D) 192
(E) 200

24 A square sheet of paper is folded twice so that a square is created again. In that square one of the corners is cut off (see the picture). Which of the pictures below cannot represent this sheet of paper once it is unfolded?

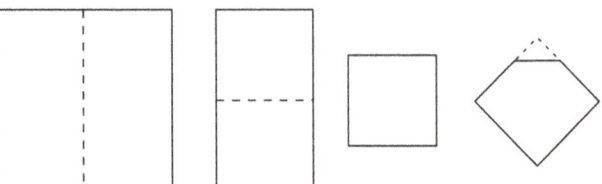

(A)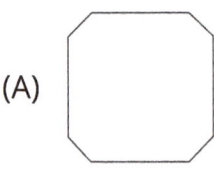

(B)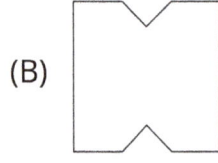

(C)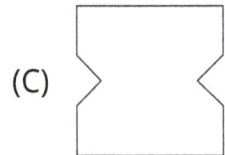

(D)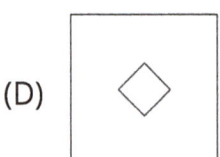

(E) Each of the pictures (A), (B), (C), (D) can represent this unfolded sheet of paper.

2009

2009

3 Points Each

1 The figure shown in the picture was made out of identical wooden cubes. How many wooden cubes were used?

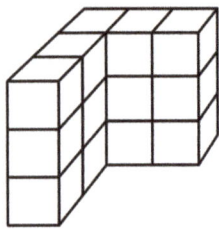

(A) 6
(B) 8
(C) 10
(D) 12
(E) 15

2 200 × 9 + 200 + 9 =

(A) 418
(B) 1909
(C) 2009
(D) 4018
(E) 20009

3 Where is the kangaroo?

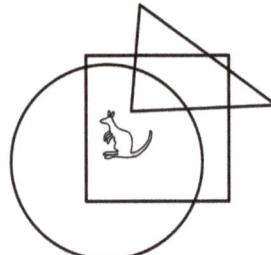

(A) In the circle and in the triangle, but not in the square.
(B) In the circle and in the square, but not in the triangle.
(C) In the triangle and in the square, but not in the circle.
(D) In the circle, but not in the square and not in the triangle.
(E) In the square, but not in the circle and not in the triangle.

4 In a certain family there are five brothers. Each one of them has one sister. How many children are in this family?

(A) 6
(B) 7
(C) 8
(D) 9
(E) 10

5 The number 930 is shown on a display (see the picture). How many little squares need to change color in order to show the number 806?

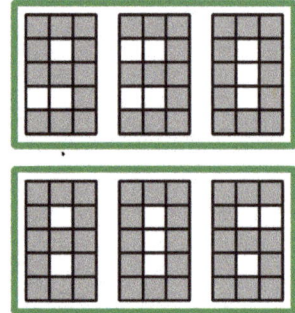

(A) 5
(B) 6
(C) 7
(D) 8
(E) 9

6 Mother bought 16 oranges. Carl ate half of them, Eva ate two, and Sophie ate the rest. How many oranges did Sophie eat?

(A) 4
(B) 6
(C) 8
(D) 10
(E) 12

7. In his garden Anthony made the path shown in the figure, using 10 tiles of size 4 m by 6 m. Anthony painted a black line between the midpoints of the tiles. How long is this black line?

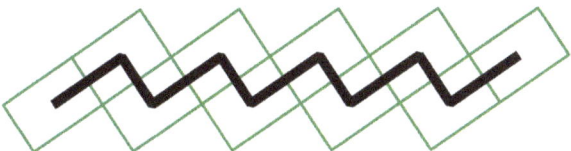

(A) 24 m
(B) 40 m
(C) 46 m
(D) 50 m
(E) 56 m

8. A certain movie is 90 minutes long. It started at 5:10 p.m. During the movie, there were two commercial breaks, one lasting 8 minutes and one lasting 5 minutes. At what time did the movie finish?

(A) at 6:13 p.m.
(B) at 6:27 p.m.
(C) at 6:47 p.m.
(D) at 6:53 p.m.
(E) at 7:13 p.m.

4 Points Each

9. A red kangaroo and a gray kangaroo together weigh 139 kilograms. The red kangaroo weighs 35 kilograms less than the gray kangaroo. How much does the gray kangaroo weigh?

(A) 104 kilograms
(B) 52 kilograms
(C) 87 kilograms
(D) 96 kilograms
(E) 53 kilograms

10. Zach was dividing a chocolate bar. He broke one row of five pieces for his brother and then one row of seven pieces for his sister, as shown in the picture. How many pieces were there in the whole chocolate bar?

(A) 28
(B) 32
(C) 35
(D) 40
(E) 54

11. A certain dance group started out with 25 boys and 19 girls. Every week 2 more boys and 3 more girls join the dance group. After how many weeks will there be the same number of boys and girls in the dance group?

(A) 6
(B) 5
(C) 4
(D) 3
(E) 2

12. A farmer has 30 cows, some chickens, and no other animals. The total number of the legs of the chickens is equal to the total number of the legs of the cows. How many animals does the farmer have altogether?

(A) 60
(B) 90
(C) 120
(D) 180
(E) 240

13. One side of a rectangle is 8 cm long, while the other is half as long. A square has the same perimeter as the rectangle. What is the length of the side of the square?

(A) 4 cm
(B) 6 cm
(C) 8 cm
(D) 12 cm
(E) 24 cm

14 Magda rolled a die four times and she obtained a total of 23 points. How many times did the roll show 6 dots?

(A) 0
(B) 1
(C) 2
(D) 3
(E) 4

15 Three squirrels, Hela, Mela, and Tola, together found 7 nuts. Each of them found a different number of nuts, but each of them found at least one nut. Hela collected the least, and Mela the most of all. How many nuts did Tola find?

(A) 1
(B) 2
(C) 3
(D) 4
(E) 5

16 Peter and Paul went to a boy scout camp. During a meeting, the scouts stood in a single row. On one side of Paul there were 27 scouts, and on the other side there were 13 scouts. Peter was standing exactly in the middle of the row. How many scouts were there between Peter and Paul?

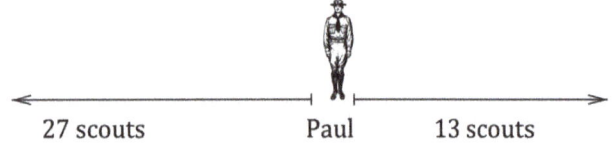

(A) 6
(B) 7
(C) 8
(D) 14
(E) 21

5 Points Each

17 Which of the figures below cannot be made using the two dominoes shown in the picture to the right?

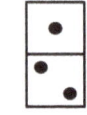 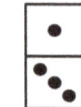

(A)
(B)
(C)
(D)
(E)

18 A secret agent wants to break a 6-digit code. He knows that the sum of the first, third, and fifth digits is equal to the sum of the second, fourth, and sixth digits. Which of the following could be the code?

(A) 8 1 * * 6 1
(B) 7 * 7 2 7 *
(C) 4 * 4 1 4 1
(D) 1 2 * 9 * 8
(E) 1 8 1 * 2 *

19 One week, Ms. Florentina sold eggs at the market every day from Monday to Friday. On Wednesday, she sold 60 eggs. On Thursday, she sold 96 eggs, and noticed that every day that week the number of eggs she sold was equal to the sum of the number of eggs she sold the two previous days. How many eggs did Ms. Florentina sell on Monday?

(A) 20
(B) 24
(C) 36
(D) 40
(E) 48

20 A certain vase contains four flowers: one red, one blue, one yellow, and one white. Kaya the Bee sat on every flower in the bouquet only once. She started with the red flower, and she did not fly directly from the yellow flower to the white flower. In how many ways could Kaya sit on all the flowers?

(A) 1
(B) 2
(C) 3
(D) 4
(E) 6

21 At 6:15 a.m. Jasper the Friendly Ghost vanished, and the crazy clock, which had been showing the right time until then, started to run at the right speed but backwards. The ghost appeared again at 7:30 p.m. that same day. What time did the crazy clock show at the moment when the ghost reappeared?

(A) 5:00 p.m.
(B) 5:45 p.m.
(C) 6:30 p.m.
(D) 7:00 p.m.
(E) 7:15 p.m.

22 The squares of a 3 × 3 table were filled in with numbers as shown in the picture. In one move, we can switch any two numbers. What is the smallest number of such moves that we need to make to get a table in which the sum of the numbers in each row is divisible by 3?

4	5	1
8	10	4
7	1	2

(A) 1
(B) 2
(C) 3
(D) 4
(E) It is impossible to get such a table.

23 Agnes was drawing figures made out of segments of length 1. At the end of each segment, she always turned at a right angle either to the left or to the right. Each time she turned right, she drew the symbol ♥ on a piece of paper, and each time she turned left, she drew the symbol ♠. One day, she drew a figure and drew these symbols in this order: ♥ ♠ ♠ ♠ ♥ ♥. Which of the following figures could Agnes have drawn?

(A)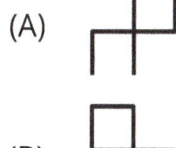

(B)

(C)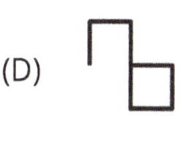

(D)

(E)

24 In the land of Funnyfeet, the left foot of each man is two sizes bigger than his right foot, and the left foot of each woman is one size bigger than her right foot. However, shoes are always sold in pairs of the same size, and only in whole sizes. A group of friends decided to buy green shoes, and to save money they bought shoes together. After they all put on the shoes that fit them, there were exactly two shoes left over, one of size 36 and another of size 45. What is the smallest possible number of people in the group?

(A) 4
(B) 5
(C) 6
(D) 7
(E) 9

2011

3 Points Each

1 Which of the numbers below is the greatest?

(A) 20 + 11
(B) 20 − 11
(C) 20 + 1 + 1
(D) 20 − 1 − 1
(E) 2 + 0 + 1 + 1

2 Michael is painting the word KANGAROO on a poster. Each day he paints one letter. He painted the first letter on a Wednesday. What day of the week will it be when he paints the last letter?

(A) Monday
(B) Tuesday
(C) Wednesday
(D) Thursday
(E) Friday

3 Which of the stones below needs to be added to the box on the right side of the scale in order for the two boxes on the scale to weigh the same? (The numbers on the stones show their weights in kilograms.)

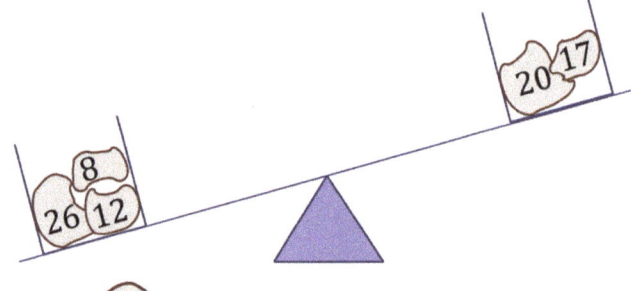

(A) 5
(B) 7
(C) 9
(D) 11
(E) 13

4 The train to Atlanta leaves three and a half hours from now. Paul got up two and a half hours ago. How many hours before the train leaves did Paul get up?

(A) two and a half
(B) three and a half
(C) four and a half
(D) five
(E) six

5 A toy is placed in one of the squares of a grid, as shown in the picture. A child moved the toy from one square to the next. He first moved it one square to the right, then one square up, then one square to the left, then one square down, and then again one square to the right. Which of the following pictures shows where the toy was in the end?

(A)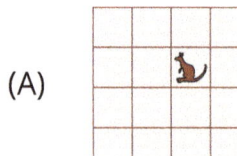

(B)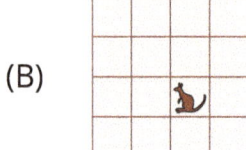

(C)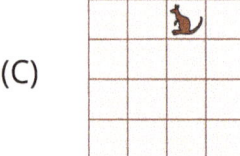

(D)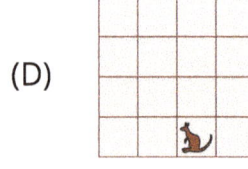

(E)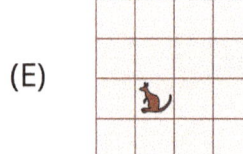

6. Ala, Lenka, and Miso went out for dessert. Lenka paid 4 dollars and 50 cents for three scoops of ice cream. Miso paid 3 dollars and 60 cents for two cookies. How much did Ala pay for one scoop of ice cream and one cookie?

(A) 3 dollars and 30 cents
(B) 4 dollars and 80 cents
(C) 5 dollars and 10 cents
(D) 6 dollars and 30 cents
(E) 8 dollars and 10 cents

7. Susan described one of the figures below. She said, "This figure is gray and it is not a rectangle." Which figure did Susan describe?

(A)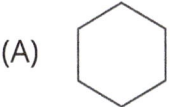

(B)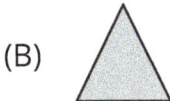

(C)

(D)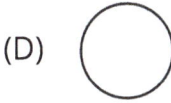

(E)

8. A tower clock strikes on the hour as many times as the number of the hour it shows (that is, it strikes 8 times at 8:00, 9 times at 9:00, and so on). It also strikes once when it is half past the hour (that is, at 8:30, 9:30, 10:30, and so on). How many times will the clock strike between 7:45 and 10:45?

(A) 6
(B) 16
(C) 27
(D) 30
(E) 33

4 Points Each

9. Which of the figures below has the greatest area?

(A)

(B)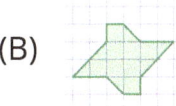

(C)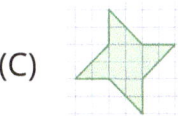

(D)

(E)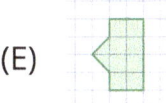

10. A farmer who raises chickens has boxes that hold 6 eggs each and boxes that hold 12 eggs each. What is the least number of boxes he needs to store 66 eggs?

(A) 5
(B) 6
(C) 9
(D) 11
(E) 13

11 Each of the students in a class drew his or her pets. Two of the students each drew a dog and a fish. Three of the students each drew a cat and a dog. The other students drew one animal each. How many students are there in this class?

(A) 11
(B) 12
(C) 13
(D) 14
(E) 17

12 During a party each of two identical cakes was divided into four equal pieces. Then each of these pieces was divided into three equal pieces. After that, each of the people at this party got a piece of cake and three pieces were left over. How many people were at this party?

(A) 13
(B) 18
(C) 21
(D) 24
(E) 27

13 The sheet is folded along the thick line. Which letter will not be covered by a gray square?

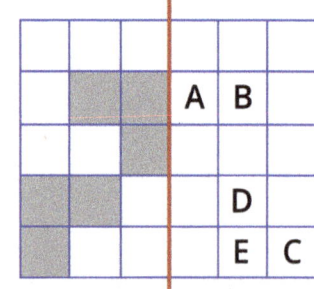

(A) A
(B) B
(C) C
(D) D
(E) E

14 There are 13 coins in John's pocket, and each of them is either a 5-cent coin or a 10-cent coin. Which of the numbers below cannot be the total value of John's coins?

(A) 60 cents
(B) 70 cents
(C) 80 cents
(D) 115 cents
(E) 125 cents

15 Ari, Chuck, Darius, Jack, Mark, and Tom were rolling a six-sided die. Each of the boys rolled the die once and each one got a different number. Ari got a result four times as great as Chuck. Darius got a result twice as great as Jack and three times as great as Mark. What number did Tom roll?

(A) 2
(B) 3
(C) 4
(D) 5
(E) 6

16 A squirrel went through the maze gathering nuts (see the picture). It could go only once through each door between the rooms of the maze. What is the greatest number of nuts it could have gathered?

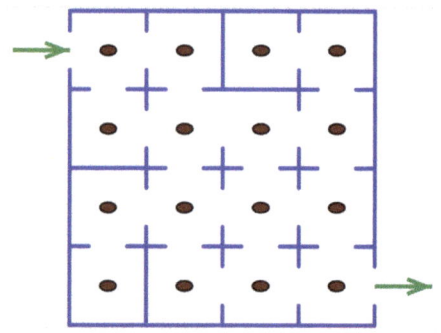

(A) 7
(B) 10
(C) 11
(D) 12
(E) 15

5 Points Each

17 A certain quiz show has the following rules: every participant has 10 points at the beginning and has to answer 10 questions. For each correct answer, the participant earns 1 point, and for each incorrect answer, the participant loses 1 point. Mrs. Smith had 14 points at the end of this quiz show. How many correct answers did she give?

(A) 7
(B) 8
(C) 9
(D) 6
(E) 4

18 Four friends, Masha, Sasha, Dasha, and Pasha, were sitting on a bench. First Masha changed places with Dasha. Then Dasha changed places with Pasha. At the end the girls sat on the bench in the following order from left to right: Masha, Sasha, Dasha, Pasha. In what order from left to right were they sitting in the beginning?

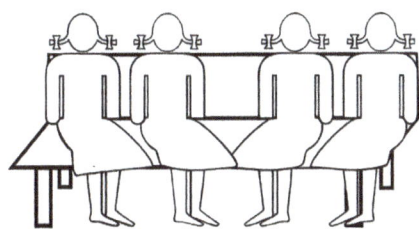

(A) Masha, Sasha, Dasha, Pasha
(B) Masha, Dasha, Pasha, Sasha
(C) Dasha, Sasha, Pasha, Masha
(D) Sasha, Masha, Dasha, Pasha
(E) Pasha, Masha, Sasha, Dasha

19 The diagram shows the distances in miles between certain towns A, B, C, D, E, and F. What is the distance in miles between towns C and D?

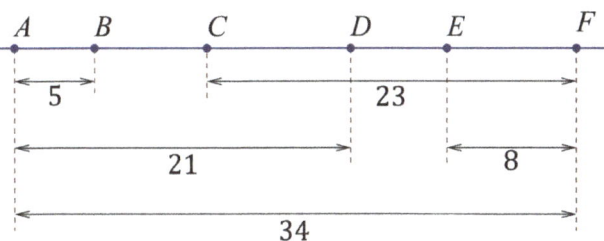

(A) 5
(B) 6
(C) 10
(D) 11
(E) 13

20 Four identical dice have been arranged in a structure as shown in the figure. The sum of the number of dots on any two opposite faces is equal to 7. What does this structure look like from behind?

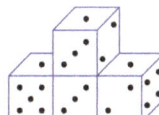

(A)

(B)

(C)

(D)

(E)

21 Olga has three cards, as shown in the picture. Using them, she can form different numbers, for example 989, 986, 866, and so on. How many different 3-digit numbers can she form using these three cards?

(A) 4
(B) 6
(C) 8
(D) 9
(E) 12

22 Andrea formed the ornament shown in the picture by using several identical pieces. The pieces cannot overlap or cover each other. Which of the following pieces could not have been used by Andrea to create the ornament?

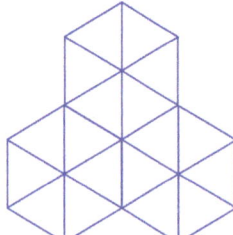

(A)

(B)

(C)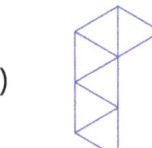

(D)

(E)

23 Picture 1 shows a castle built out of identical cubes. When you look at the same castle from above it looks as shown in Picture 2. Every cube lies either on the ground or on the top of another cube. How many cubes were used to build the castle?

Picture 1.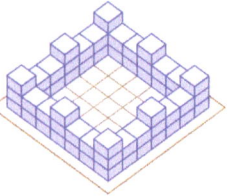

Picture 2.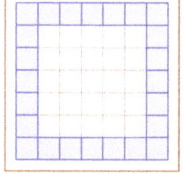

(A) 56
(B) 60
(C) 64
(D) 68
(E) 72

24 Chris wrote the numbers 6 and 7 in the circles as shown in the picture. He will then write each of the numbers 1, 2, 3, 4, 5, and 8 in the circles so that the sum of the numbers on each of the sides of the square is equal to 13. What will the sum of the numbers in the circles shaded yellow be?

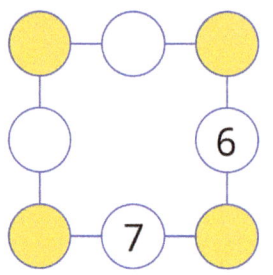

(A) 12
(B) 13
(C) 14
(D) 15
(E) 16

2013

3 Points Each

1 In which figure is the number of black kangaroos larger than the number of white kangaroos?

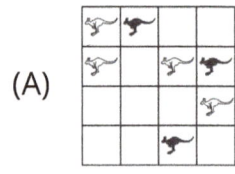

(A)

(B)

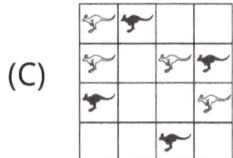

(C)

(D)

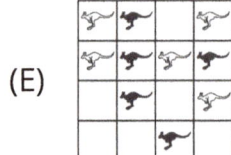

(E)

---

2 Aline writes a correct calculation. Then she covers two digits that are the same with stickers (see the picture). Which digit is under the stickers?

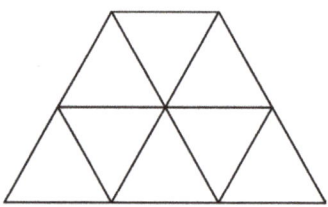

(A) 2
(B) 4
(C) 5
(D) 7
(E) 8

---

3 In what way should the last four circles be shaded so that the pattern is continued?

(A)
(B)
(C)
(D)
(E)

---

4 How many triangles can be seen in the picture?

(A) 9
(B) 10
(C) 11
(D) 13
(E) 12

5. In the London 2012 Olympics, USA won the most medals: 46 gold, 29 silver, and 29 bronze. China won the second most medals with 38 gold, 27 silver, and 23 bronze medals. How many more medals than China did USA win?

(A) 6
(B) 14
(C) 16
(D) 24
(E) 26

6. Daniel had a package of 36 pieces of candy. Without breaking any pieces of candy, he divided all the candy equally among his friends. Which of the following was definitely not the number of his friends?

(A) 2
(B) 3
(C) 4
(D) 5
(E) 6

7. Vero's mom prepares sandwiches with two slices of bread each. A package of bread has 24 slices. How many sandwiches can she prepare from two and a half packages of bread?

(A) 24
(B) 30
(C) 48
(D) 34
(E) 26

8. About the number 325, five boys said:
Andy: "This is a 3-digit number."
Barry: "All the digits are different."
Charlie: "The sum of the digits is 10."
Danny: "The ones digit is 5."
Eddie: "All the digits are odd."
Which of the boys was wrong?

(A) Andy
(B) Barry
(C) Charlie
(D) Danny
(E) Eddie

4 Points Each

9. A rectangular mirror was broken. Which of the following pieces is missing from the picture of the broken mirror?

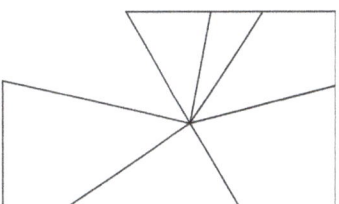

(A)

(B)

(C)

(D)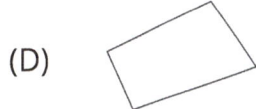

(E)

10 Each time Pinocchio lies, his nose gets 6 cm longer. Each time he tells the truth, his nose gets 2 cm shorter. After his nose was 9 cm long, he told three lies and made two true statements. How long was Pinocchio's nose afterwards?

(A) 14 cm
(B) 15 cm
(C) 19 cm
(D) 23 cm
(E) 31 cm

11 In a shop you can buy oranges in boxes of three different sizes: boxes of 5 oranges, boxes of 9 oranges, or boxes of 10 oranges. Pedro wants to buy exactly 48 oranges. What is the smallest number of boxes he can buy?

(A) 8
(B) 7
(C) 6
(D) 5
(E) 4

12 Ann starts walking in the direction of the arrow. At every intersection of the streets she turns either to the right or to the left. First she goes to the right, then to the left, then again to the left, then to the right, then to the left, and finally again to the left. Then Ann is finally walking towards

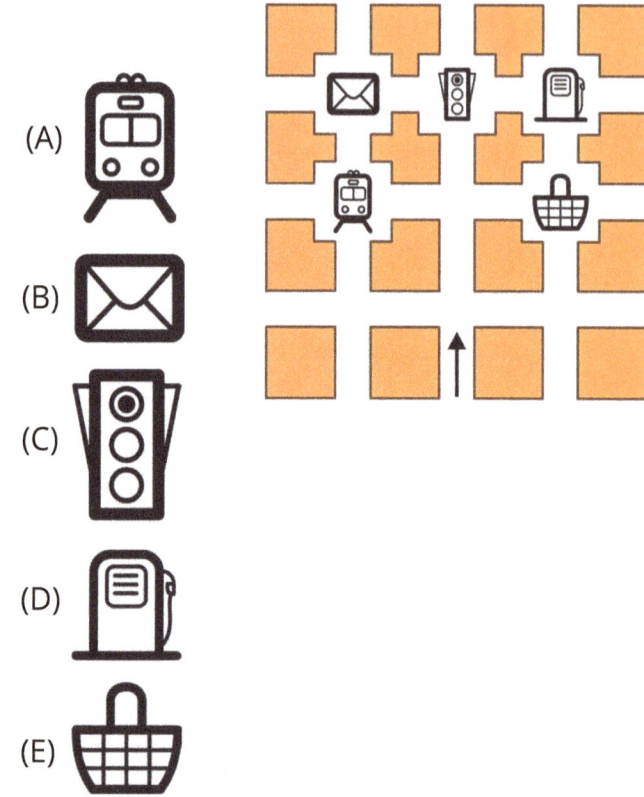

(A) train
(B) envelope
(C) traffic light
(D) gas pump
(E) basket

13 Classmates Andy, Betty, Cathie, and Dannie were born in the same year. Their birthdays were on February 20th, April 12th, May 12th, and May 25th, not necessarily in this order. Betty and Andy were born in the same month. Andy and Cathie were born on the same day of different months. Who of these classmates is the oldest?

(A) Andy
(B) Betty
(C) Cathie
(D) Dannie
(E) It is impossible to determine.

14 At Adventure Park, each of 30 children took part in at least one of two events. 15 of them took part in the "moving bridge" contest, and 20 of them went down the zipline. How many of the children took part in both events?

(A) 25
(B) 15
(C) 30
(D) 10
(E) 5

15 Which of the following pieces fits with the piece in the picture so that together they form a rectangle?

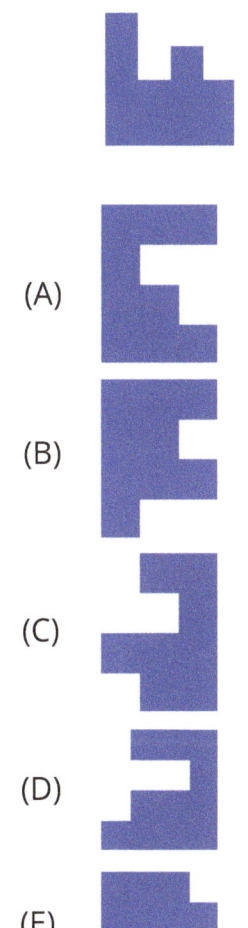

16 The number 35 has the property that it is divisible by the digit in the ones position, because 35 divided by 5 is 7. The number 38 does not have this property. How many numbers greater than 21 and smaller than 30 have this property?

(A) 2
(B) 3
(C) 4
(D) 5
(E) 6

5 Points Each

17 Joining the midpoints of the sides of the triangle in the drawing, we obtain a smaller triangle. We repeat this one more time with the smaller triangle. How many triangles of the same size as the smallest resulting triangle fit in the original drawing?

(A) 5
(B) 8
(C) 10
(D) 16
(E) 32

18 After January 1st, 2013, how many years will pass before the following event happens for the first time: the product of the digits in the year is greater than the sum of these digits?

(A) 87
(B) 98
(C) 101
(D) 102
(E) 103

19 In December Tosha-the-Cat slept for exactly 3 weeks. How many minutes did she stay awake during this month?

(A) (31 − 7) × 3 × 24 × 60
(B) (31 − 7 × 3) × 24 × 60
(C) (30 − 7 × 3) × 24 × 60
(D) (31 − 7) × 24 × 60
(E) (31 − 7 × 3) × 24 × 60 × 60

20 Basil has several domino tiles, as shown below. He wants to arrange them in a line according to the following "domino rule": in any two neighboring tiles, the neighboring squares must have the same number of dots. What is the largest number of tiles he can arrange in this way?

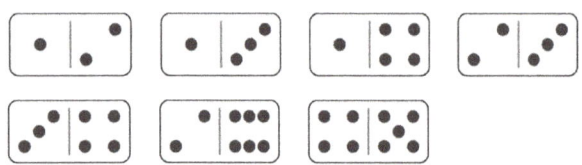

(A) 3
(B) 4
(C) 5
(D) 6
(E) 7

21 Cristi has to sell 10 glass bells which vary in price: 1 dollar, 2 dollars, 3 dollars, 4 dollars, 5 dollars, 6 dollars, 7 dollars, 8 dollars, 9 dollars, and 10 dollars. In how many ways can Cristi divide all the glass bells into three packages so that each of the packages has the same price?

(A) 1
(B) 2
(C) 3
(D) 4
(E) Such division is not possible.

22 Peter bought a rug 36 in wide and 60 in long. The rug has a pattern of small squares containing either a sun or a moon, as can be seen in the figure. You can see that along the width there are 9 squares. When the rug is fully unrolled, how many moons can be seen?

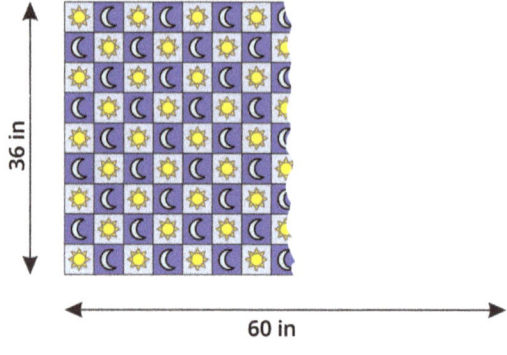

(A) 68
(B) 67
(C) 65
(D) 63
(E) 60

23 Baby Roo wrote down as few numbers as possible using only the digits 0 and 1 to get 2013 as the sum. How many numbers did Baby Roo write?

(A) 2
(B) 3
(C) 4
(D) 5
(E) 204

24 Beatrice has many pieces like the gray one in the picture. At least how many of these gray pieces are needed to make a completely full gray square?

(A) 3
(B) 4
(C) 6
(D) 8
(E) 16

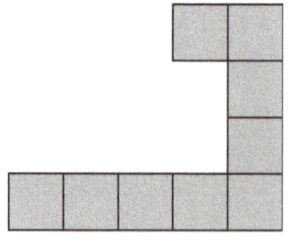

3 Points Each

1.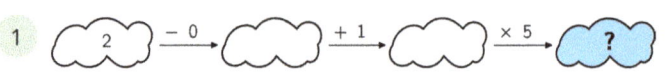

(A) 6
(B) 7
(C) 8
(D) 10
(E) 15

2. Eric has 10 identical metal strips.

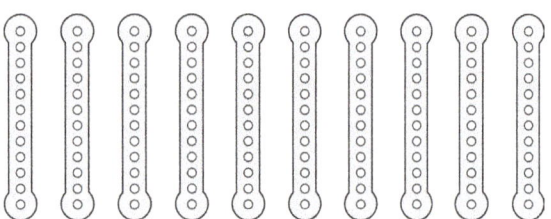

He used screws to connect pairs of them together into five long strips.

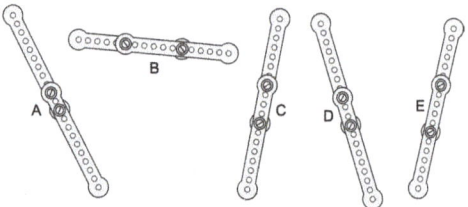

Which strip is the longest?

(A) A
(B) B
(C) C
(D) D
(E) E

3. Which number is hidden behind the square in the equation to the right?

(A) 2
(B) 3
(C) 4
(D) 5
(E) 6

4. Which of the expressions below has the greatest value?

(A) (1000 − 100) ÷ 10
(B) (1000 − 10) ÷ 9
(C) (1000 − 1) ÷ 9
(D) (1000 − 100) ÷ 9
(E) (1000 − 10) ÷ 10

5. We start drawing segments connecting every other dot on the circle until we are back at the number 1. The first two segments are already drawn, as shown in the picture to the right. Which figure do we get?

(A)

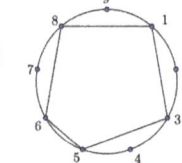

(B)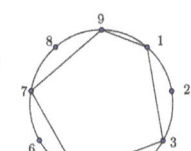

(C)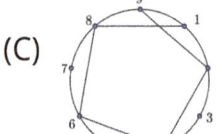

(D)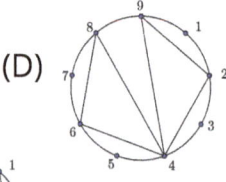

(E)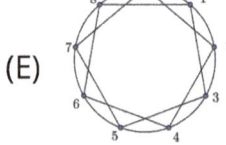

6. A certain whole number has two digits. The product of the digits of this number is 15. The sum of the digits of this number is:

(A) 2
(B) 4
(C) 6
(D) 7
(E) 8

7. In the picture, we see an island with a highly indented coastline and several frogs. How many of these frogs are sitting on the island?

(A) 5
(B) 6
(C) 7
(D) 8
(E) 9

8. This year, March 19 falls on a Thursday. What day of the week will it be in 30 days?

(A) Wednesday
(B) Thursday
(C) Friday
(D) Saturday
(E) Sunday

4 Points Each

9. My umbrella has KANGAROO written on top. It is shown in the picture on the right. Which of the pictures below also shows my umbrella?

(A)

(B)

(C)

(D)

(E)

10. Basil wants to cut the shape shown in Figure 1 into identical triangles as shown in Figure 2. How many triangles will he get?

(A) 8
(B) 12
(C) 14
(D) 15
(E) 16

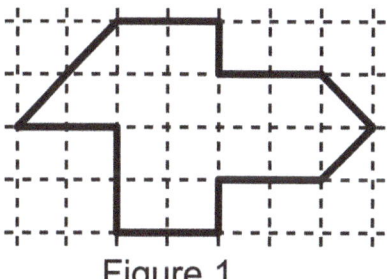

Figure 1

Figure 2

11 Luis had 7 apples and 2 bananas. He gave 2 apples to Yuri, who in return gave some bananas to Luis. Now Luis has as many apples as bananas. How many bananas did Yuri give to Luis?

(A) 2
(B) 3
(C) 4
(D) 5
(E) 7

12 Grandma bought some candy. She gave each of her grandchildren 4 pieces of candy and had 2 pieces left. If she wanted to give each of them 5 pieces of candy, she would be 2 pieces short. How many grandchildren does she have?

(A) 3
(B) 4
(C) 5
(D) 6
(E) 7

13 In a speed skating competition, 10 skaters reached the finish line. The number of skaters who came in before Tom was 3 less than the number of skaters who came in after him. Which place did Tom end up in?

(A) 1
(B) 3
(C) 4
(D) 6
(E) 7

14 Josip has 4 toys: a car, an airplane, a ball, and a ship. He wants to put them all in a row on a shelf. Both the ship and the airplane have to be next to the car. In how many ways can he arrange the toys so that this condition is fulfilled?

(A) 2
(B) 4
(C) 5
(D) 6
(E) 8

15 Pete rides a bicycle in a park that has paths as shown in the picture. He starts from point S and goes in the direction of the arrow. At the first crossroad he turns right, then at the next crossroad he turns left, then right again, then left again, and so on, in that order. Which letter will he not pass?

(A) A
(B) B
(C) C
(D) D
(E) E

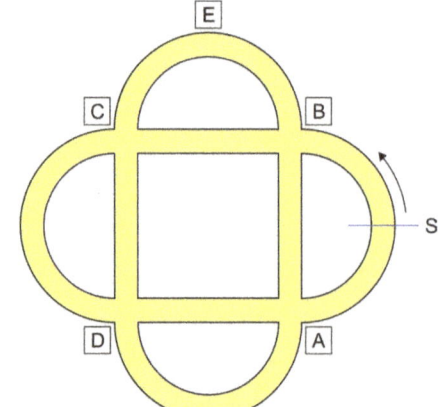

16 There are 5 ladybugs. Two ladybugs are friends with each other only if the numbers of spots that they have differ exactly by 1. On Kangaroo Day each of the ladybugs sent one text greeting to each of her friends. How many text messages were sent?

(A) 2
(B) 4
(C) 6
(D) 8
(E) 9

5 Points Each

17 The figure is divided into three identical pieces. What does each of the pieces look like?

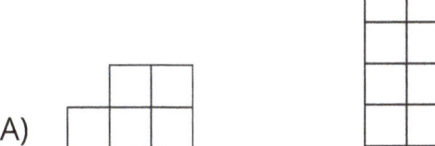

(A)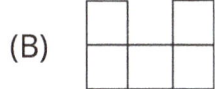

(B)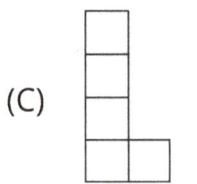

(C)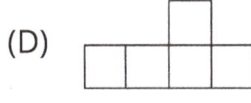

(D)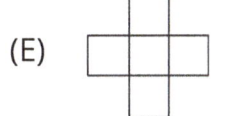

(E)

18 Jack built a cube using 27 small cubes which are colored either gray or white. No two of the small cubes which are the same color have a common face. How many white cubes did Jack use?

(A) 9
(B) 12
(C) 13
(D) 14
(E) 17

19 We can fill a certain barrel with water if we use water from 6 small pitchers, 3 medium pitchers, and 1 large pitcher, or from 2 small pitchers, 1 medium pitcher, and 3 large pitchers. If we use only large pitchers of water, how many of them do we need to fill the barrel?

(A) 4
(B) 5
(C) 6
(D) 7
(E) 8

20 The numbers 2, 3, 5, 6, and 7 are written in the squares of the cross in such a way that the sum of the numbers in the row is equal to the sum of the numbers in the column. Which of the numbers can be written in the center square of the cross?

(A) only 3
(B) only 5
(C) only 7
(D) 5 or 7
(E) 3, 5, or 7

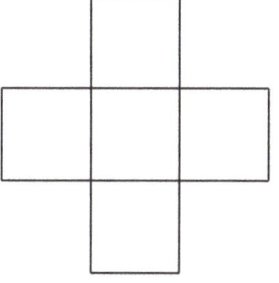

21 Peter has ten balls, numbered from 0 to 9. He gave four of the balls to George and three to Ann. Then each of the three friends multiplied the numbers on their balls. As the result, Peter got 0, George got 72, and Ann got 90. What is the sum of the numbers on the balls that Peter kept for himself?

(A) 11
(B) 12
(C) 13
(D) 14
(E) 15

22 Three ropes are laid down on the floor as shown on the right. You can make one big, complete rope by adding one of the sets of rope ends shown in the pictures below (without changing their positions). Which of the sets will make one complete rope?

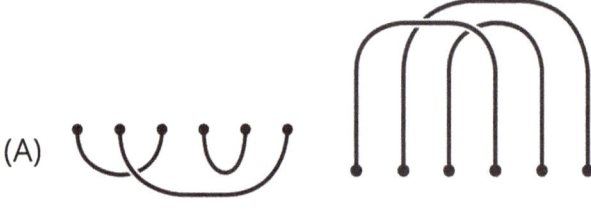

(A)

(B)

(C)

(D)

(E)

23 We have three transparent sheets with patterns. We can rotate the sheets, but not turn them over. Then we put all three sheets exactly one on top of the other. What is the maximum possible number of black squares seen in the square obtained in this way if we look at it from above?

(A) 5
(B) 6
(C) 7
(D) 8
(E) 9

24 Anna, Berta, Charlie, David, and Elisa were baking cookies on Friday and Saturday. Over the two days, Anna made 24 cookies, Berta 25, Charlie 26, David 27, and Elisa 28. Over the two days, one of them made twice as many cookies as on Friday, one 3 times as many, one 4 times as many, one 5 times as many, and one 6 times as many. Who baked the most cookies on Friday?

(A) Anna
(B) Berta
(C) Charlie
(D) David
(E) Elisa

2017

2017

3 Points Each

1. Which of the pieces (A) through (E) will fit between the two pieces shown below so the two equations are true?

(A)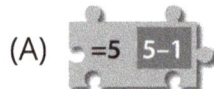

(B)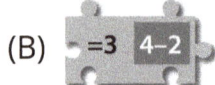

(C)

(D)

(E)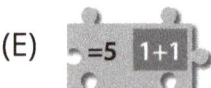

2. John looks out the window. He sees half of the kangaroos in the park. How many kangaroos are there in the park?

(A) 12
(B) 14
(C) 16
(D) 18
(E) 20

3. Two transparent grids have some dark squares, as shown. They both slide into place on top of the board shown in the middle. Now the pictures behind the dark squares cannot be seen. Only one of the pictures can still be seen. Which one is it?

(A)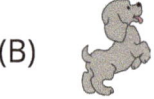

(B)

(C)

(D)

(E)

4. A picture of footprints was turned upside down. Which set of footprints is missing?

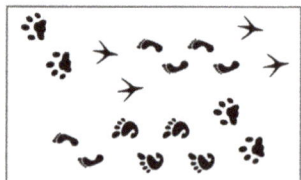 

(A)

(B)

(C)

(D)

(E)

5. What number is hidden behind the panda?

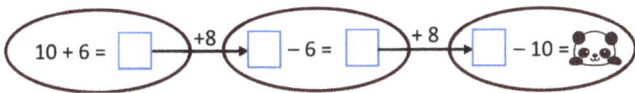

(A) 16
(B) 18
(C) 20
(D) 24
(E) 28

6. The table shows correct sums. What number is in the box with the question mark?

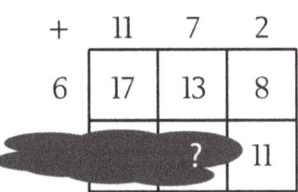

(A) 10
(B) 12
(C) 13
(D) 15
(E) 16

7. Dolly accidentally broke the mirror into pieces. How many pieces have exactly four sides?

(A) 2
(B) 3
(C) 4
(D) 5
(E) 6

8. Here is a necklace with six beads.

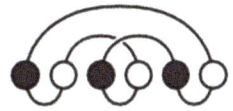

Which of the pictures below shows the same necklace?

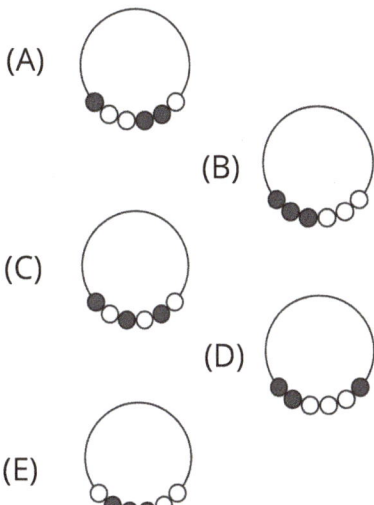

(A)
(B)
(C)
(D)
(E)

4 Points Each

9. The picture shows the front of Ann's house. The back of her house has three windows and no door. Which view does Ann see when she looks at the back of her house?

(A)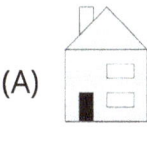

(B)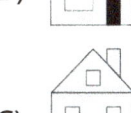

(C)

(D)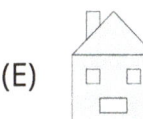

(E)

10.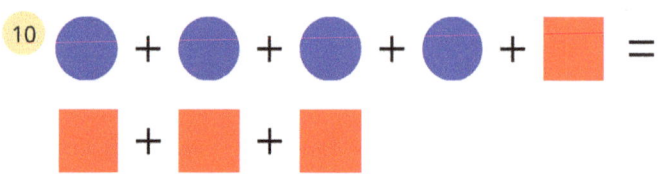

Which of the following is true?

(A)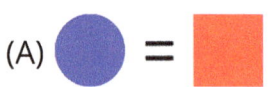

(B)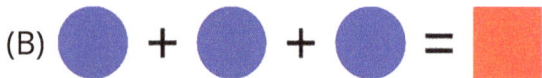

(C)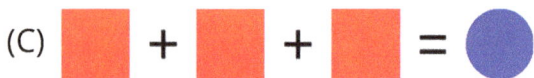

(D)

(E)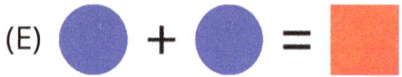

11. Balloons are sold in packets of 5, 10, and 25. Marius buys exactly 70 balloons. What is the smallest number of packets he can buy?

(A) 3
(B) 4
(C) 5
(D) 6
(E) 7

12. Bob folded a piece of paper. He cut exactly one hole in the paper. Then he unfolded the piece of paper and saw the result as shown in the picture to the right. How did Bob fold his piece of paper?

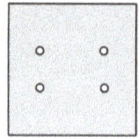

(A)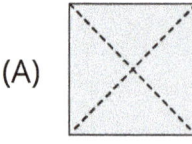

(B)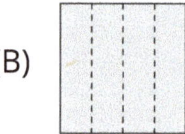

(C)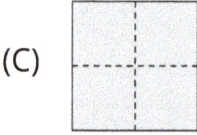

(D)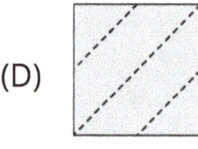

(E)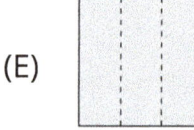

13. There is a tournament at the pool. First, 13 children signed up, and then another 19 children signed up. Six teams with an equal number of members each are needed for the tournament. At least how many more children need to sign up so that the six teams can be formed?

(A) 1
(B) 2
(C) 3
(D) 4
(E) 5

14 Numbers are placed in the cells of the 4 × 4 square shown in the picture. Mary finds the 2 × 2 square where the sum of the numbers in the four cells is the largest. What is that sum?

1	2	1	3
4	1	1	2
1	7	3	2
2	1	3	1

(A) 11
(B) 12
(C) 13
(D) 14
(E) 15

---

15 David wants to prepare a meal with 5 dishes using a stove with only 2 burners. The times needed to cook the 5 dishes are 40 minutes, 15 minutes, 35 minutes, 10 minutes, and 45 minutes. What is the shortest time in which he can do it? (He may only remove a dish from the stove when it is done cooking.)

(A) 60 minutes
(B) 70 minutes
(C) 75 minutes
(D) 80 minutes
(E) 85 minutes

---

16 Which number should be written in the circle with the question mark?

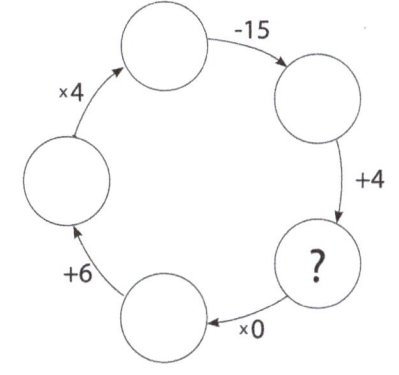

(A) 10
(B) 11
(C) 12
(D) 13
(E) 14

---

5 Points Each

17 The picture shows a group of building blocks and a plan of the same group. Some ink spilled on the plan. What is the sum of the numbers under the ink spills?

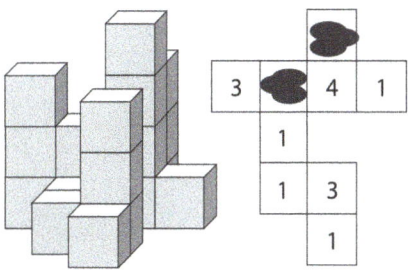

(A) 3
(B) 4
(C) 5
(D) 6
(E) 7

---

18 How long is the train?

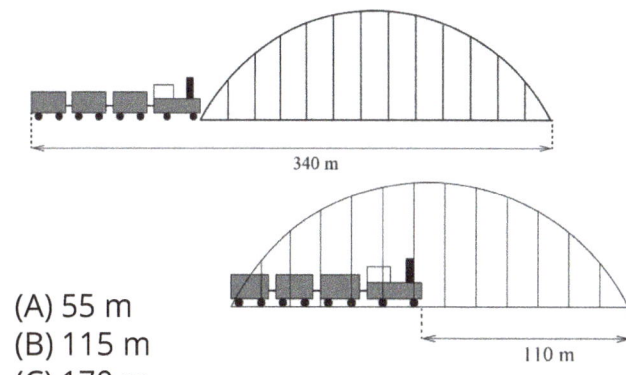

(A) 55 m
(B) 115 m
(C) 170 m
(D) 220 m
(E) 230 m

---

19 The ancient Romans used Roman numerals. We still use them today. I=1, V=5, X=10, L=50, C=100, D=500, M=1000. John was born in February of the year MMVII. How old was John on March 16, 2017?

(A) I
(B) III
(C) IV
(D) V
(E) X

20 A small zoo has a giraffe, an elephant, a lion, and a turtle. Susan wants to plan a tour where she sees 2 different animals. She does not want to start with the lion. How many different tours can she plan?

(A) 3
(B) 7
(C) 8
(D) 9
(E) 12

21 Four brothers ate 11 cookies in total. Each of them ate at least one cookie and no two of them ate the same number of cookies. Three of them ate 9 cookies in total and one of them ate exactly 3 cookies. How many cookies did the boy who ate the largest number of cookies eat?

(A) 3
(B) 4
(C) 5
(D) 6
(E) 7

22 Zosia hid a smiley ☺ in some of the cells of the table. In some of the other cells she wrote the number of smileys in the neighboring cells as shown in the picture. Two cells are neighboring if they share a common side or a common corner. How many smileys did she hide?

	3	3	
2			
		2	
	1		

(A) 4
(B) 5
(C) 7
(D) 8
(E) 11

23 Each of ten bags contains a different number of pieces of candy. The number of pieces of candy in each bag ranges from 1 to 10. Each of five boys took two bags of candy. Alex got 5 pieces of candy, Bob got 7 pieces, Charles got 9 pieces, and Dennis got 15 pieces. How many pieces of candy did Eric get?

(A) 9
(B) 11
(C) 13
(D) 17
(E) 19

24 Kate has 4 flowers, one with 6 petals, one with 7 petals, one with 8 petals, and one with 11 petals. Kate tears off one petal from three flowers. She does this several times, choosing any three flowers each time. She stops when she can no longer tear one petal from three flowers. What is the smallest number of petals which can remain?

(A) 1
(B) 2
(C) 3
(D) 4
(E) 5

2019

2019

3 Points Each

1. The higher the step on the podium, the higher the rank of the runner. Who finished third?

(A) A
(B) B
(C) C
(D) D
(E) E

2. In the pictures, each dot stands for 1 and each bar stands for 5.

For example, 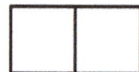 stands for 8. Which picture stands for 12?

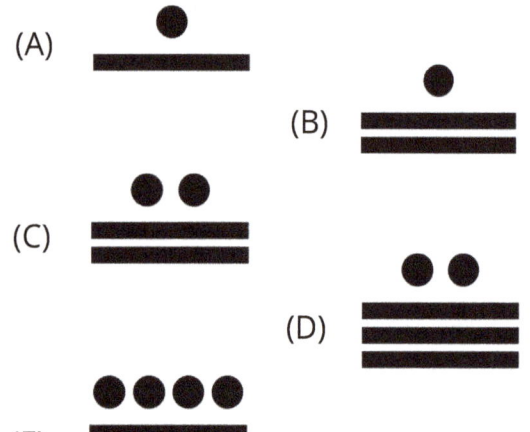

3. Yesterday was Sunday. What day is tomorrow?

(A) Tuesday
(B) Thursday
(C) Wednesday
(D) Monday
(E) Saturday

4. There are two holes in the cover of a book. When the book is open, it looks like this:

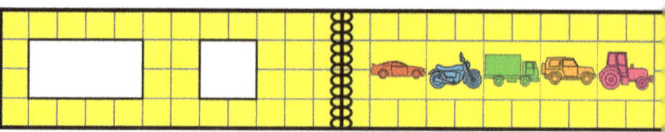

Which pictures does Olaf see through the holes when he closes the book?

(A)

(B)

(C)

(D)

(E)

5. Karina cuts out one piece like this

from the sheet shown to the right. Which piece can she get?

(A) (star, club)

(B) (star, spade)

(C)

(D)

(E)

6. Three people walked across a field of snow wearing muddy shoes. In which order did they do this?

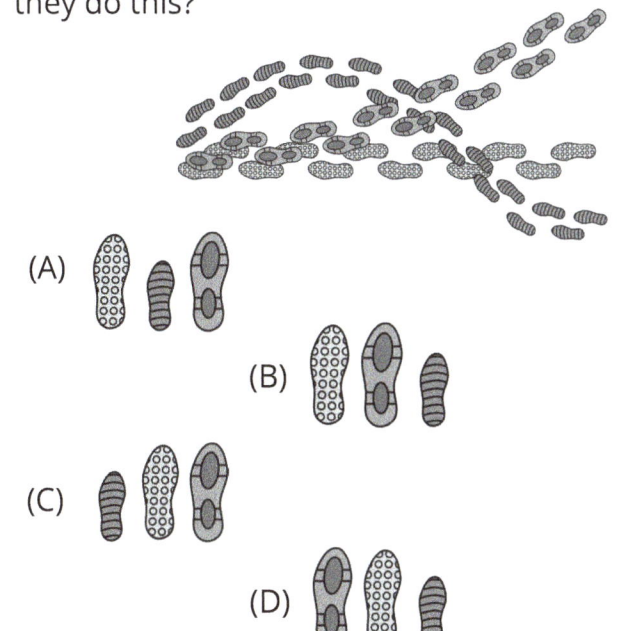

(A)
(B)
(C)
(D)
(E)

7. Pia is making shapes using the connected sticks shown in the picture.

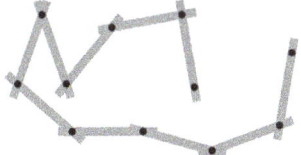

Which of the following shapes uses more sticks than Pia has?

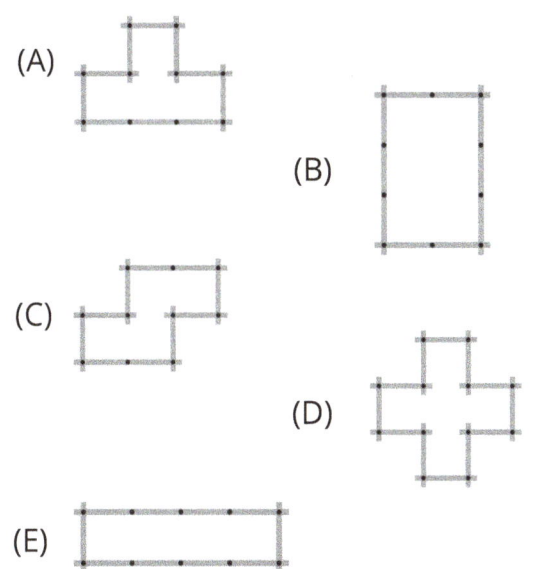

(A)
(B)
(C)
(D)
(E)

8. What number should replace the question mark when all the calculations are completed correctly?

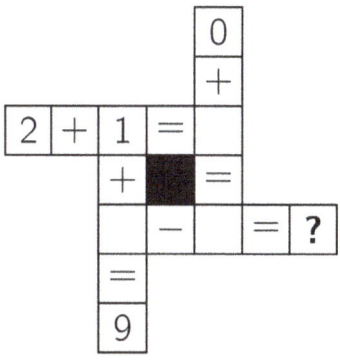

(A) 4
(B) 5
(C) 6
(D) 7
(E) 8

4 Points Each

9. Linda pinned up 3 photos in a row on a cork board using 8 pins. Peter wants to pin up 7 photos in the same way. How many pins does he need?

(A) 14
(B) 16
(C) 18
(D) 22
(E) 26

10. Dennis wants to remove one cell from the shape:

How many of the shapes below can he get?

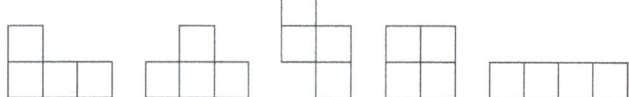

(A) 1
(B) 2
(C) 3
(D) 4
(E) 5

11. Six strips are woven into a pattern as shown:

What does the pattern look like from the back?

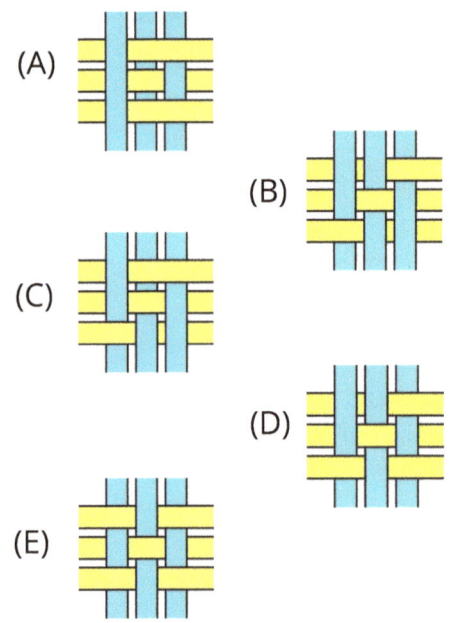

(A)
(B)
(C)
(D)
(E)

12. The weight of the toy dog is a whole number. How much does one toy dog weigh?

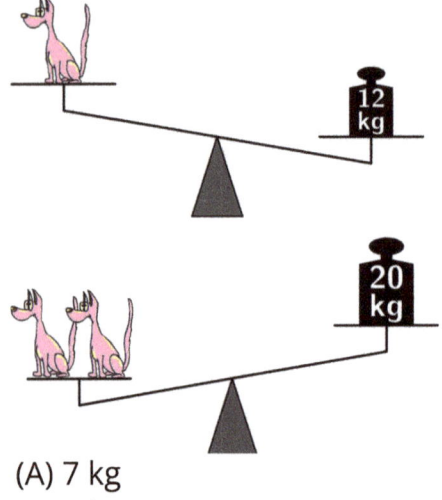

(A) 7 kg
(B) 8 kg
(C) 9 kg
(D) 10 kg
(E) 11 kg

13. Sara has 16 blue marbles. She can trade marbles in two ways: 3 blue marbles for 1 red marble, and 2 red marbles for 5 green marbles. What is the maximum number of green marbles she can get?

(A) 5
(B) 10
(C) 13
(D) 15
(E) 20

14. Steven wants to write each of the digits 2, 0, 1, and 9 in one of the boxes of the sum:

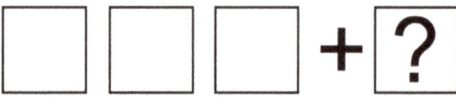

He wants to get the largest possible answer. Which digit can he write instead of the question mark?

(A) Either 0 or 1
(B) Either 0 or 2
(C) Only 0
(D) Only 1
(E) Only 2

15. A glass full of water weighs 400 grams. An empty glass weighs 100 grams. How many grams does a glass half-full with water weigh?

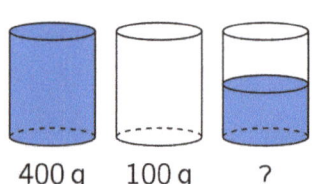

400 g   100 g   ?

(A) 150
(B) 200
(C) 225
(D) 250
(E) 300

16.  Together we cost 5 cents.    Together we cost 7 cents.

 Together we cost 10 cents.    How much do we cost together?

(A) 8 cents
(B) 9 cents
(C) 10 cents
(D) 11 cents
(E) 12 cents

5 Points Each

17. Each shape stands for a different number. The sum of the three numbers in each row is shown to the right of the row.

Which number does the stand for?

(A) 2
(B) 3
(C) 4
(D) 5
(E) 6

18. Anna used 32 small white squares to frame a 7 by 7 picture. How many of these small white squares does she need to frame a 10 by 10 picture?

(A) 36
(B) 40
(C) 44
(D) 48
(E) 52

19. The pages of a book are numbered 1, 2, 3, 4, 5, and so on. The digit 5 appears exactly 16 times. What is the maximum number of pages this book can have?

(A) 49
(B) 64
(C) 66
(D) 74
(E) 80

20. A hallway has the dimensions shown in the picture. A cat walks on the dashed line along the middle of the hallway. How many meters does the cat walk?

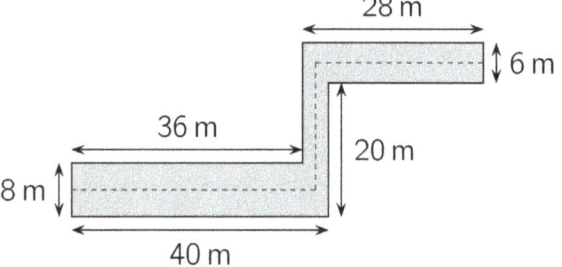

(A) 63
(B) 68
(C) 69
(D) 71
(E) 83

21. In a park there are 15 animals: cows, cats, and kangaroos. We know that precisely 10 are not cows and precisely 8 are not cats. How many kangaroos are there in the park?

(A) 2
(B) 3
(C) 4
(D) 8
(E) 10

22 Mary has 9 small triangles: 3 of them are red (R), 3 are yellow (Y), and 3 are blue (B). She wants to form a big triangle by putting together these 9 small triangles so that any two triangles with an edge in common are different colors. Mary places some small triangles as shown in the picture. Which of the following statements is true after she has finished?

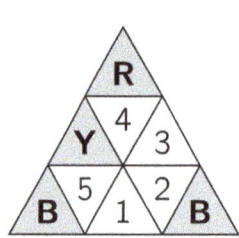

(A) 1 is yellow and 3 is red
(B) 1 is blue and 2 is red
(C) 1 and 3 are red
(D) 5 is red and 2 is yellow
(E) 1 and 3 are yellow

23 There are five children: Alek, Bartek, Czarek, Darek, and Edek. One of them ate a cookie.

Alek says: "I did not eat the cookie."
Bartek says: "I ate the cookie."
Czarek says: "Edek did not eat the cookie."
Darek says: "I did not eat the cookie."
Edek says: "Alek ate the cookie."

Only one child is lying. Who ate the cookie?

(A) Alek
(B) Bartek
(C) Czarek
(D) Darek
(E) Edek

24 Emil started to hang up towels using two pegs for each towel as shown in figure 1. He realized that he would not have enough pegs and began to hang up the towels as shown in figure 2. Altogether, he hung up 35 towels and used 58 pegs. How many towels did Emil hang up in the way shown in figure 1?

figure 1
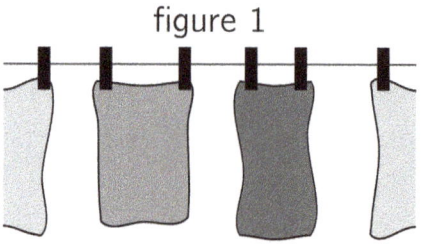

figure 2
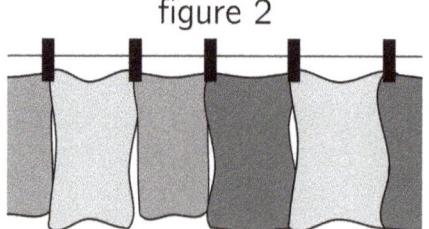

(A) 12
(B) 13
(C) 21
(D) 22
(E) 23

2021

2021

3 Points Each

1. Erik has 4 bricks:

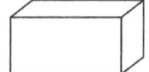 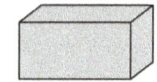  

Which of the cubes shown below can he make with his 4 bricks?

(A)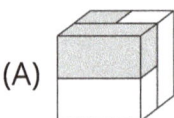

(B)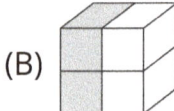

(C)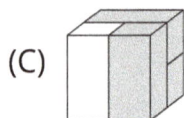

(D)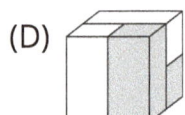

(E)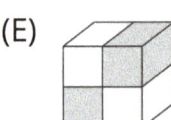

2. How many fish will have their heads pointing towards the ring when we straighten the line?

(A) 3
(B) 5
(C) 6
(D) 7
(E) 8

3. When you put the 4 puzzle pieces together correctly, they form a rectangle with a calculation on it. What is the result of this calculation?

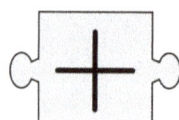

(A) 6
(B) 15
(C) 18
(D) 24
(E) 33

4. Alaya drew a picture of the sun. Which of the following answers is part of her picture?

(A)

(B)

(C)

(D)

(E)

5. Five boys competed in a shooting challenge. Ricky scored the most points. Which target was Ricky's?

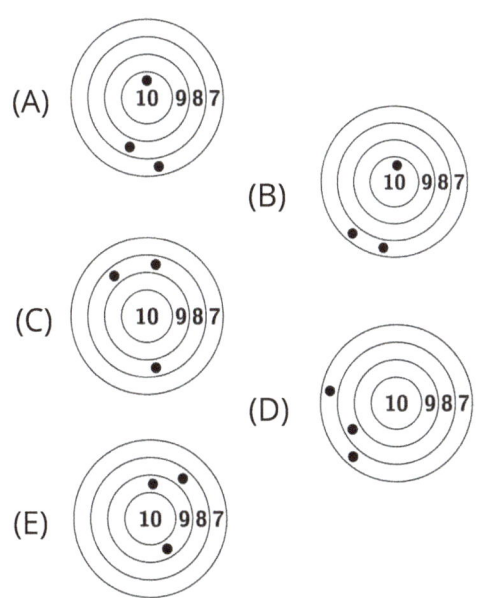

6. A measuring tape is wrapped around a cylinder. Which number should be at the place shown by the question mark?

(A) 33
(B) 42
(C) 48
(D) 53
(E) 69

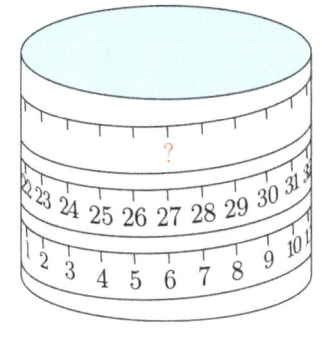

7. Denise fired a silver rocket and a gold rocket at the same time. The rockets exploded into 20 stars in total. The gold rocket exploded into 6 more stars than the silver one. How many stars did the gold rocket explode into?

(A) 9
(B) 10
(C) 12
(D) 13
(E) 15

8. Rosana has some balls of 3 different colors. Balls of the same color have the same weight. What is the weight of each white ball?

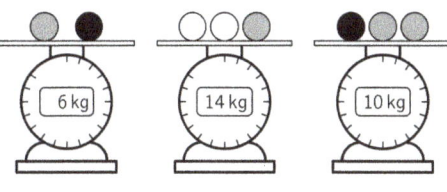

(A) 3 kg
(B) 4 kg
(C) 5 kg
(D) 6 kg
(E) 7 kg

4 Points Each

9. Nisa has 3 different types of cards in a game:

apple

cherry

and grapes

They can be arranged in sets of 5. In each set, Nisa can choose just 2 cards and swap their places. She wants to arrange the cards so that all the cards with the same fruit are next to each other. For which set is this not possible?

(A)

(B)

(C)

(D)

(E)

10 Sofie wants to pick 5 different shapes from the containers. She can only pick 1 shape from each container. Which shape must she pick from container 4?

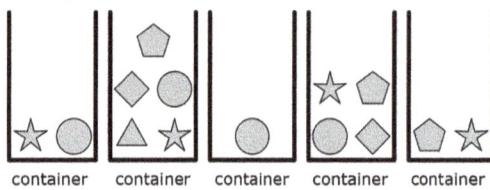

(A)

(B)

(C)

(D)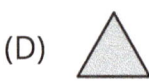

(E)

11 18 cubes are colored white, gray, or black and are arranged as shown.

The figures show the white and the black parts.

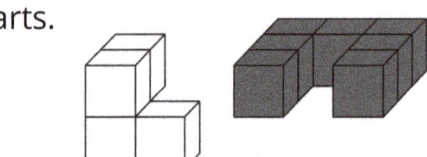

Which of the following is the gray part?

(A)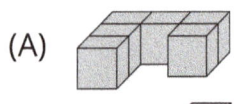

(B)

(C)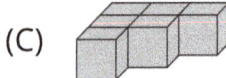

(D)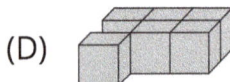

(E)

12 The 5 balls shown begin to move simultaneously in the directions indicated by their arrows.

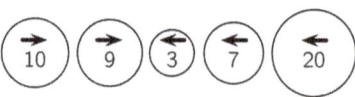

When two balls going in opposite directions collide, the bigger ball swallows the smaller one and increases its value by the value of the smaller ball. The bigger ball continues to move in its original direction, as shown in the following example.

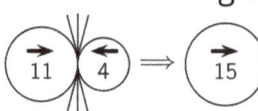

What is the final result of the collisions of the 5 balls shown?

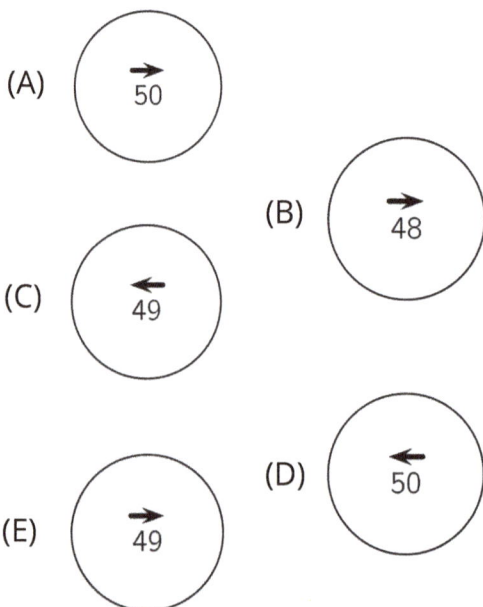

13 In an ice cream shop there was some money in a drawer. After selling 6 ice cream cones, there are 70 dollars in the drawer. After selling a total of 16 ice cream cones, there are 120 dollars in the drawer. How many dollars were there in the drawer at the start?

(A) 20
(B) 30
(C) 40
(D) 50
(E) 60

14 The Koala ate some leaves from 3 branches. Each branch had 20 leaves. The Koala ate a few leaves from the first branch and then ate as many leaves from the second branch as were left on the first branch. Then it ate 2 leaves from the third branch. How many leaves in total were left on the 3 branches?

(A) 20
(B) 22
(C) 28
(D) 32
(E) 38

15 On a tall building there are 4 fire escape ladders, as shown. The heights of 3 ladders are at their tops. What is the height of the shortest ladder?

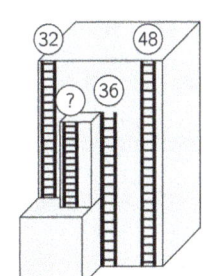

(A) 12
(B) 14
(C) 16
(D) 20
(E) 22

16 Nora is playing with 3 cups on the kitchen table. In each move, she takes the cup on the left, flips it over, and puts it to the right of the other cups. The picture shows the first move. What do the cups look like after 10 moves?

(A)
(B)
(C)
(D)
(E)

5 Points Each

17 Eva has the 5 stickers shown:

She placed each of them on the 5 squares of this board

| 1 | 2 | 3 | 4 | 5 |

one on each square, so that is not on square 5, is on square 1, and is next to and .

On which square did Eva place ?

(A) 1
(B) 2
(C) 3
(D) 4
(E) 5

18 7 cards are arranged as shown. Each card has two numbers on it, with one of them written upside down. The teacher wants to rearrange the cards so that the sum of the numbers in the top row is the same as the sum of the numbers in the bottom row. She can do this by turning one of the cards upside down. Which card must she turn?

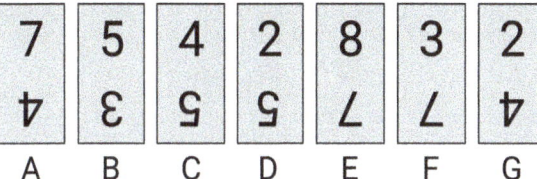

(A) A
(B) C
(C) D
(D) F
(E) G

19 The numbers 1 to 9 are placed in the squares shown, with a number in each square. The sums of all pairs of neighboring numbers are shown. Which number is in the shaded square?

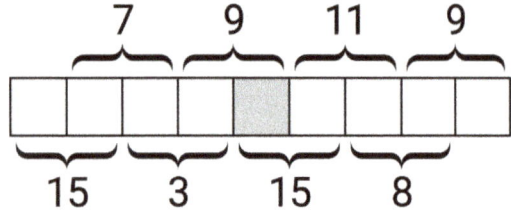

(A) 4
(B) 5
(C) 6
(D) 7
(E) 8

20 Mia throws darts at balloons worth 3, 9, 13, 14, and 18 points. She scores 30 points in total. Which balloon does Mia definitely hit?

(A) 3
(B) 9
(C) 13
(D) 14
(E) 18

21 A box has fewer than 50 cookies in it. The cookies can be divided evenly between 2, 3, or 4 children. However, they cannot be divided evenly between 7 children because 6 more cookies would be needed. How many cookies are there in the box?

(A) 12
(B) 24
(C) 30
(D) 36
(E) 48

22 Each of the 5 boxes contains either apples or bananas, but not both. The total weight of all the bananas is 3 times the weight of all the apples. Which boxes contain apples?

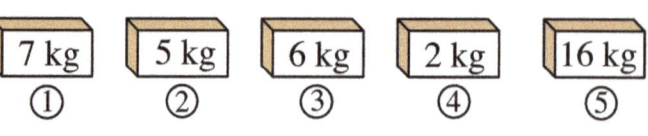

(A) 1 and 2
(B) 2 and 3
(C) 2 and 4
(D) 3 and 4
(E) 1 and 4

23 Elena wants to write the numbers from 1 to 9 in the squares shown. The arrows always point from a smaller number to a larger one. She has already written 5 and 7. Which number should she write instead of the question mark?

(A) 2
(B) 3
(C) 4
(D) 6
(E) 8

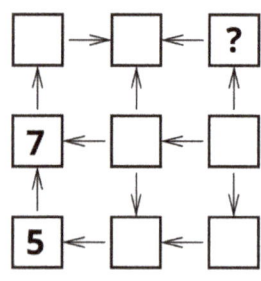

24 Martin placed 3 different types of objects, hexagons ⬡, squares ■, and triangles △ on sets of scales, as shown. What does he need to put on the left-hand side on the third set of scales for these scales to balance?

(A) 1 square
(B) 2 squares
(C) 1 hexagon
(D) 1 triangle
(E) 2 triangles

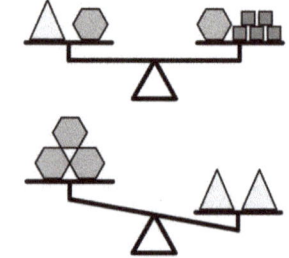

2023

3 Points Each

1. Zoe lit 5 identical candles all at the same time. They stopped burning at different times and now look as shown in the picture. Which candle stopped burning first?

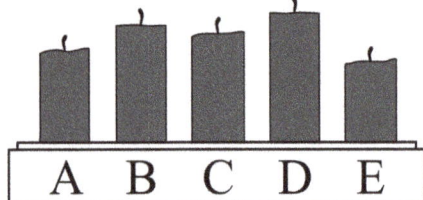

(A) A
(B) B
(C) C
(D) D
(E) E

2. The 2 kangaroo coins with the question marks on them have the same value. What is this value?

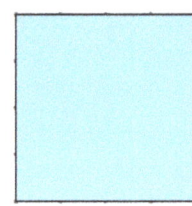

(A) 1
(B) 2
(C) 5
(D) 10
(E) 20

3. A gray circle with 2 large holes in it is put on top of a clock face.

 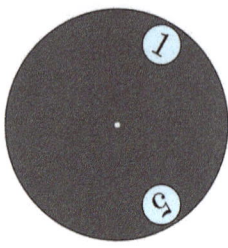

The gray circle is turned around its center. Which 2 numbers is it possible to see at the same time?

(A) 4 and 9
(B) 5 and 9
(C) 5 and 10
(D) 6 and 9
(E) 7 and 12

4. Alice has these puzzle pieces:

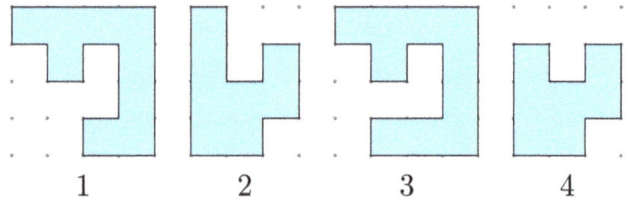

Which 2 pieces can she put together to form this square?

(A) 1 and 2
(B) 1 and 3
(C) 1 and 4
(D) 2 and 3
(E) 2 and 4

5. A light engineer in the theater turns the lights on and off. She uses the plan shown. How long in total are exactly 2 of the lights on at the same time?

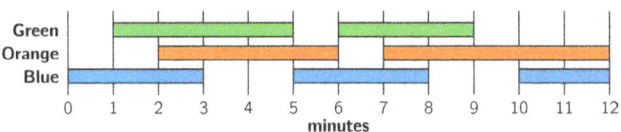

(A) 2 minutes
(B) 6 minutes
(C) 8 minutes
(D) 9 minutes
(E) 10 minutes

6. Aarav folds the transparent paper along the dashed line. What can he then see?

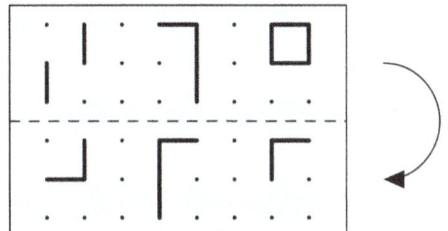

(A)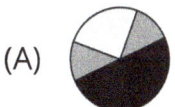

(B)

(C)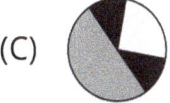

(D)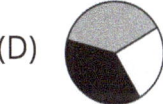

(E)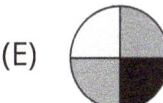

7. Anna has 4 discs of different sizes. She wants to build a tower of 3 discs so that each disc is smaller than the disc below it. How many different towers can Anna make?

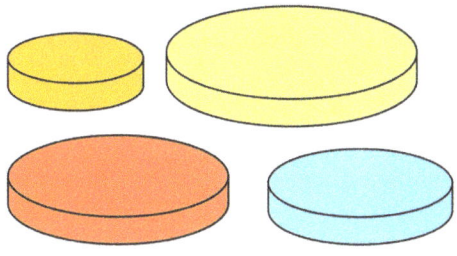

(A) 1
(B) 2
(C) 4
(D) 5
(E) 6

8. Danny glued the 2 pieces of paper without cutting them, on top of the black circle. What can he not make?

4 Points Each

9 The shape on the right is covered with the 5 pieces below. Which piece will cover the dot?

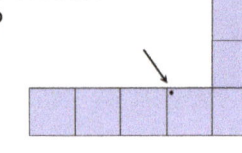

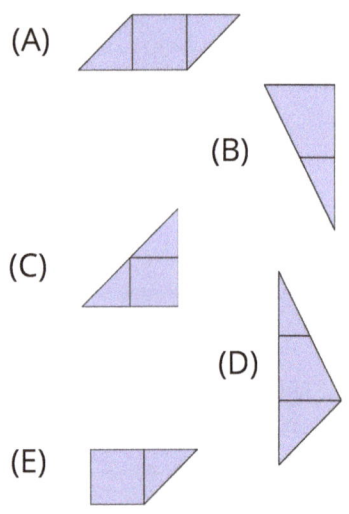

(A)

(B)

(C)

(D)

(E)

10 There are six weights: 1 kg, 2 kg, 3 kg, 4 kg, 5 kg, and 6 kg. Amy puts five of them on the scales and puts one weight aside. The scales balance. Which weight did she put aside?

(A) 1 kg
(B) 2 kg
(C) 3 kg
(D) 4 kg
(E) We can't be sure.

11 Andrew has a 60 cm ruler. Unfortunately, some of the markings have faded. He can still measure the lengths 10 cm, 20 cm, 30 cm, 40 cm, 50 cm, and 60 cm using his ruler only once. Which is his ruler?

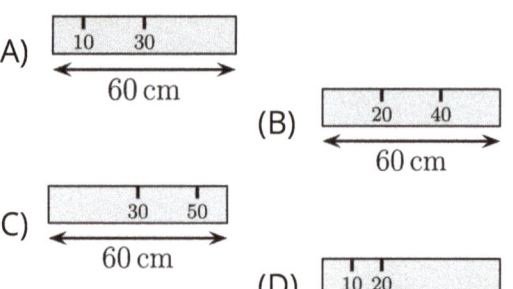

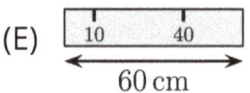

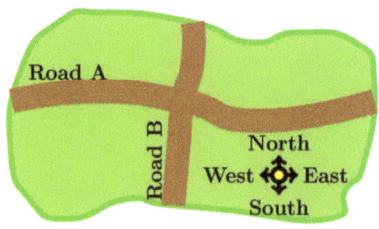

12 There are 7 houses north of Road A, 8 houses east of Road B, and 5 houses south of Road A. How many houses are west of Road B?

(A) 4
(B) 5
(C) 6
(D) 7
(E) 8

13 There are 8 cars waiting in a line for the ferry. Every car contains either 2 or 3 people. There are 19 people in total waiting for the ferry. How many cars contain exactly 2 people?

(A) 2
(B) 3
(C) 4
(D) 5
(E) 6

14 The Metro line has 6 stations, A, B, C, D, E, and F. The train stops at every station. When it reaches one of the two end stations, it changes its direction. The train engineer started driving at station B and her first stop was station C. Which station will be her 96th stop?

(A) A
(B) B
(C) C
(D) D
(E) E

15 Aanya wants to paint the circles in the picture. She wants to paint any 2 circles connected with a line different colors. What is the smallest number of colors she needs?

(A) 2
(B) 3
(C) 4
(D) 5
(E) 6

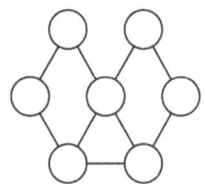

16 Sam walks through the two-story maze from the entrance to the exit, passing 3 wall stickers. In what order will she see them?

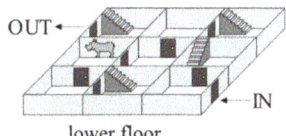 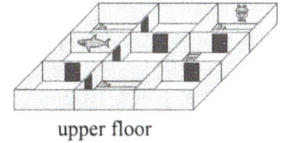
lower floor   upper floor

(A)

(B)

(C)

(D)

(E)

5 Points Each

17 6 beavers and 2 kangaroos are standing in a line. Among any 3 consecutively numbered animals, exactly 1 is a kangaroo. Which of the following numbered animals is a kangaroo?

(A) 1
(B) 2
(C) 3
(D) 4
(E) 5

18 Rebecca folds a square piece of paper twice. Then she cuts off one corner. Next, she unfolds the paper. What does the paper look like once unfolded?

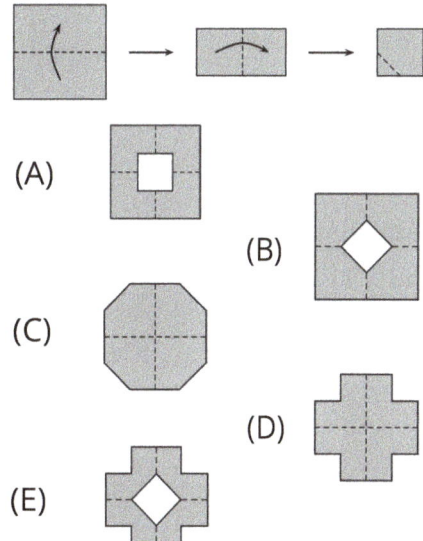

19 Hermione, Harry, and Ron always walk into the common room one at a time. Hermione is never first, Harry is never second, and Ron is never third. In how many different orders could they walk in?

(A) 1
(B) 2
(C) 3
(D) 4
(E) 6

20. There are five clocks on the wall. It is known that one clock is an hour fast, one clock is an hour slow, one clock shows the correct time, and two clocks have stopped. Which clock shows the correct time?

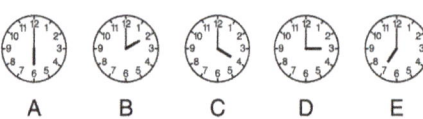

(A) A
(B) B
(C) C
(D) D
(E) E

21. Adam and Brenda have 9 marbles each. Together, they have 8 red and 10 blue marbles. Brenda has twice as many blue marbles as red marbles. How many blue marbles does Adam have?

(A) 3
(B) 4
(C) 5
(D) 6
(E) 0

22. Chloe has two machines. When she puts a square sheet of paper in machine R, it turns the paper 90° clockwise, as shown in the picture. When she puts the paper in machine S, it stamps the paper with a ♣.

In which order are the machines used to produce the result shown?

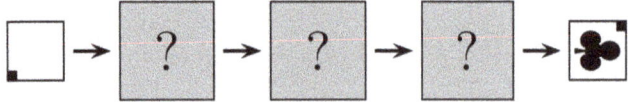

(A) SRR
(B) RSR
(C) RSS
(D) RRS
(E) SRS

23. Teacher Claire wants to write the numbers 1 to 7 in the circles. Inside each circle she writes one number. She wants the sum of the numbers in two neighboring circles to match the number shown on the connecting line. What number must she write inside the green circle?

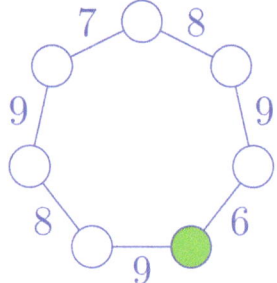

(A) 1
(B) 2
(C) 3
(D) 4
(E) 5

24. Maria has shaded exactly 5 cells in a 4 × 4 grid. She challenges 5 of her friends to guess which cells she has shaded. The grids they have drawn are shown below. Maria looks at them and says: "One of you is right and each of the rest of you has four cells correct." Which is the correct answer?

(A)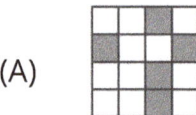

(B)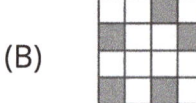

(C)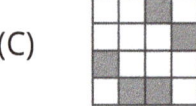

(D)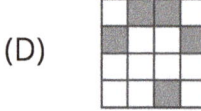

(E)

Part II
Solutions

1999

3 Point Solutions

1 (C) 10
Beata's fruits are: three apples, two oranges, and five peaches, so Beata has 3 + 2 + 5 = 10 pieces of fruit.

2 (E) 6
The picture has the circles colored to make them easier to see. The part common to all four circles is shaded.

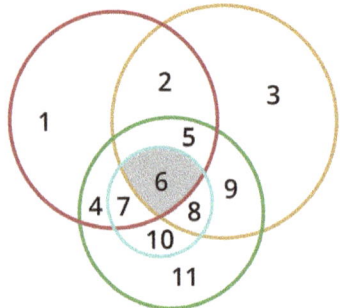

3 (B) 4
To get 5 pieces, you need to break the stick in only 4 places as shown in the picture below.

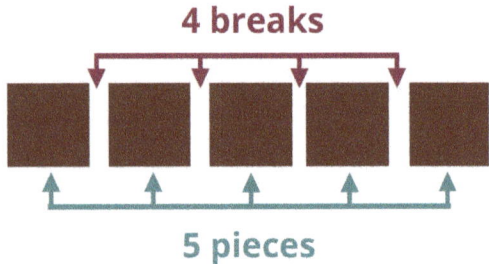

4 (C) 4
Karl is 10 – 3 = 7 years older than Alice. Because the difference of their ages is 7 years, he will be twice her age when she is 7, and he is 14. That will be 14 – 10 = 4 (or 7 – 3 = 4) years from now.

5 (D) Both ways have the same length.
The matching distances are marked by arrows of the same color.

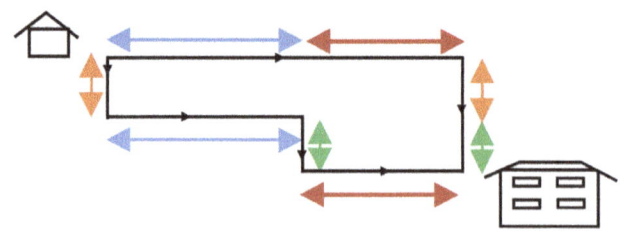

The matching shows that both ways have the same length.

6 (E) 6
We can split the boys into 4 groups, each of the same size as the group of girls. Thus, the class of 30 students consists of 5 equal groups and the number of girls is one-fifth of 30, which is 6.

7 (C) 155 g
Since the scales must balance,
5 g + 200 g = orange + 50 g
5 g + 150 g + 50 g = orange + 50 g
155 g = orange
The orange weighs 155 g.

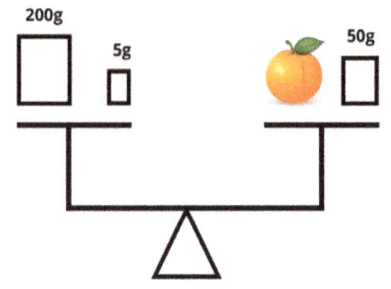

8 (B) 24 cm
A whole is made up of two halves, so the 12 cm is half the length the tail. Hence, the tail is 12 cm + 12 cm = 24 cm long.

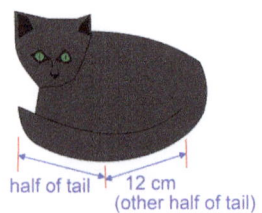

4 Point Solutions

9 **(E) Saturday**
It is also a Sunday 8 weeks after mom's birthday. 8 weeks have 8 × 7 = 56 days and 56 − 1 = 55, so dad's birthday is one day before Sunday, which is Saturday.

10 **(B) 7**
6 games are not always enough since each team could win exactly 3 games. With 7 games one team must have at least 4 wins and the other team less than 4 wins.

11 **(C) 54**
Subtracting instead of adding is moving in the opposite direction on a number line. Instead moving to the right John moved 27 units to the left. The distance between John's result and the result he should have gotten is 27 + 27 = 54.

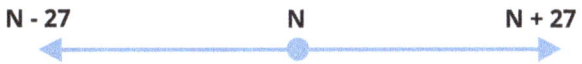

12 **(A) 64**
Look at the top layer of the big cube. It is made up of 4 × 4 small cubes since 4 cm = 4 × 1 cm. There are 3 other layers below the top layer, so there are altogether 4 layers of 4 × 4 small cubes. The big cube contains 4 × 4 × 4 = 64 small cubes.

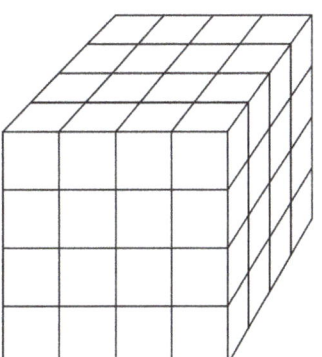

13 **(D) 1 kilogram**
The pail plus half of the milk weighs 13 kilograms. If we add the other half of the milk to the pail, then the pail is filled with milk to the top and the total weight is 25 kilograms. Hence, half of the milk weighs 25 kilograms − 13 kilograms = 12 kilograms. All the milk weighs 2 × 12 kilograms = 24 kilograms, so the pail weighs 25 kilograms − 24 kilograms = 1 kilogram.

14 **(A) 50 g**
Tom eats 5 × 6 g = 30 g of jam from the jar each day. After 20 days he will eat 20 × 30 g = 600 g of jam, so 650 g − 600 g = 50 g of jam will be left in the jar.

15 **(D) 1331**
The kangaroo has 11 children, and 11 × 11 = (10 + 1) × 11 = 10 × 11 + 11 = 110 + 11 = 121 grandchildren. Each grandchild has 11 children, so the kangaroo has 11 × 121 = (10 + 1) × 121 = 10 × 121 + 121 = 1210 + 121 = 1331 great-grandchildren.

16 **(D) 4**
Each Kowalski child has a sister and a brother, so there is a girl and a boy in the Kowalski family. The boy has a brother, so there are at least two boys in the family. The girl has a sister, so there are at least two girls in the family. Thus, the least possible number of children in the family is 4.

5 Point Solutions

17 **(D) 110**
The page numbers are 10 and 11 since they are consecutive numbers that add up to 21. The product of the two page numbers is 10 × 11 = 110.

18 **(C) 4**
143 = 40 + 40 + 40 + 23. 3 cows can feed 40 + 40 + 40 children and we need at least one 1 more cow to feed the other 23 children, so Father Virgil needs at least 4 cows.

19 **(C) 126**

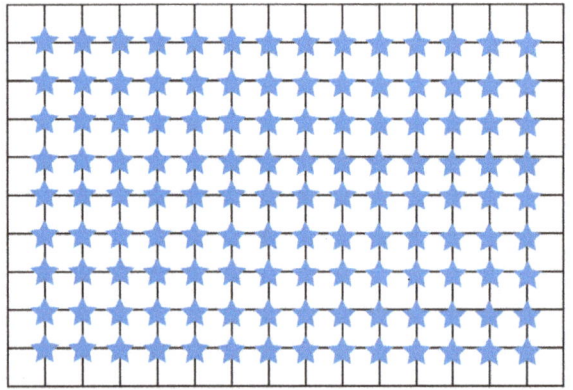

1 m = 10 × 10 cm and 1.5 m = 15 × 10 cm, so the bedspread is a rectangle made out of 10 × 15 square scraps. There are 9 rows of buttons (9 = 10 − 1) and there are 14 buttons (14 = 15 − 1) in each row, so the number of all buttons needed is 9 × 14 = 90 + 36 = 126.

20 **(A) 192 cm**
The length of Pinocchio's nose is:

2 × 3 cm after the 1st lie,
2 × 2 × 3 cm after the 2nd lie,
2 × 2 × 2 × 3 cm after the 3rd lie,
2 × 2 × 2 × 2 × 3 cm after the 4th lie,
2 × 2 × 2 × 2 × 2 × 3 cm after the 5th lie, and
2 × 2 × 2 × 2 × 2 × 2 × 3 cm after the 6th lie.

Consecutive products of 2 are: 2, 4, 8, 16, 32, and 64, so after 6 lies Pinocchio's nose is 64 × 3 cm = 192 cm long.

21 **(A) 18**
1 pig has as many legs as a duck and a chicken together, so 72 legs (half of the 144 legs) belong to the pigs. Each pig has 4 legs, so there are 18 pigs since 4 × 18 = 72. Hence, there are 18 ducks in the yard since the number of ducks is the same as the number of pigs.

22 **(D) 450**
The differences are: 501 − 51 = 450, 502 − 52 = 450, 503 − 53 = 450, 504 − 54 = 450, and 505 − 55 = 450. Whichever number was chosen, the difference is always 450.

23 (C) 8

All the grandchildren except one received 10 pieces of candy each. After that Grandma wants each grandchild to have the candy. She can do it if every grandchild has 8 pieces of candy and there are still 6 pieces left over. Every grandchild with 10 pieces of candy must give up 2 pieces of candy, so Grandma can collect 8 pieces for the one grandchild without candy and keep 6 extra pieces. Together she would collect 8 + 6 = 14 pieces, which means that there were 7 grandchildren who originally had 10 pieces each, because 2 × 7 = 14. Including the grandchild originally without candy, she has 8 grandchildren.

24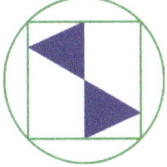

At 2:00 p.m. the figure looks exactly the same as at 12:00 p.m. since any number of full rotations brings it back to how it was at first. Now, we need to see how much it rotates in the 15 minutes after 2 o'clock. Since 15 minutes is ¼ of an hour, the figure will rotate ¼ of a full circle, or 90°, clockwise. The figures below have lines added to help visualize the rotation better. The orange figure in the second picture shows the figure after the rotation.

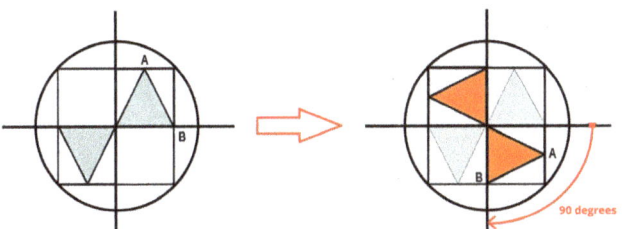

2001

2001

3 Point Solutions

1 **(C) 3 hr**
Each candle burns for 3 hours. Julia lit only 2 candles at first, so there were 2 candles that were not lit at all before the window was closed. Since these two candles were "brand new" when lit after the window was closed, it will take 3 hours for these candles to burn out.

2 **(D) 8**
Joseph keeps 6 original unbroken sticks and adds 2 new sticks by breaking the seventh stick into two pieces, so he has 8 sticks now.

3 **(D) 400 grams**

Count all the full squares, (2 + 4 + 5 + 5) × 2 = 32 counted by columns. There are 16 half-squares which is equal to 8 full squares. The area of the heart is 32 + 8 = 40 squares altogether. Thus, the weight of the whole heart is 40 × 10 grams = 400 grams.

4 **(C) 20**
Altogether there were 12 × 10 = 120 pairs of shoes in the store. The centipedes bought 3 × 30 + 2 × 5 = 90 + 10 = 100 pairs. Thus, there were 120 – 100 = 20 pairs of shoes left.

5 **(B) 31**
Kaya gave her mother one cup of berries on the first day. Each day Kaya doubles how many cups of berries she gives her mother, so Kaya gave her mother 1 + 2 + 4 + 8 + 16 = 31 cups of berries over 5 days.

6 **(E) 18 − 6 ÷ 3 = 16**
Remember to perform the operations in the correct order (multiplication and division first, from left to right, then addition and subtraction from left to right). Also remember that you have to perform the operations within parentheses first.

Calculations:

(A) 12 ÷ (4 + 8) = 12 ÷ 12 = 1 ≠ 11
(B) 8 × 2 + 3 = 16 + 3 = 19 ≠ 40
(C) 2 × 3 + 4 × 5 = 6 + 20 = 26 ≠ 50
(D) (10 + 8) ÷ 2 = 18 ÷ 2 = 9 ≠ 14
(E) 18 − 6 ÷ 3 = 18 − 2 = 16

7 **(E) 5**
Together there are 19 + 12 = 31 children. The smallest multiple of 6 greater than 31 is 36. 36 − 31 = 5. So, 5 students need to join them.

8 **(E) 8 cm**
Altogether, the length of all 4 sticks is 4 × 14 cm = 56 cm. The total space between them is 80 cm − 56 cm = 24 cm. There are 3 gaps between the sticks, so each distance has to be 24 cm ÷ 3 = 8 cm.

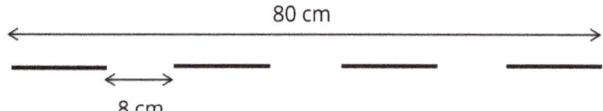

4 Point Solutions

9 **(C) 3**
Abby will be twice Bobby's age when Bobby will be as old as the difference between their ages, which is 3. This will happen 3 years later, when Abby is 6 and Bobby is 3.

10 **(D) 640 m**
The depth of this cave is 221 m + 419 m = 640 m.

11 **(E) 5**
The 6 smallest and different numbers are 1, 2, 3, 4, 5, and 6. Their sum is 21 which exceeds 20, so you cannot distribute 20 pieces of candy according to these rules among 6 children. If you ignore the number 1, you can follow the rules and distribute 20 pieces of candy among 5 children; 2, 3, 4, 5, 6 are five different numbers and 2 + 3 + 4 + 5 + 6 = 20. So, the maximum number of children who received candy is 5.

12 **(E)**
Each figure is made up of 5 blocks in the shape of the letter L and one more cube. Only in (E) that one cube has a common edge with the corner cube of the piece shown.

13 **(C) 50**
By adding 17 and 34 we get 51 cars. The car in which both girls sat in was counted twice, so there were 51 − 1 = 50 cars.

14 **(B) 3 times**
Adam and Bart have the same number of chestnuts at the beginning and Adam can split his group of chestnuts into two halves, so Bart can do the same and together they have 4 halves of chestnuts. If Adam gives Bart one half of his chestnuts, then Adam will have 1 half and Bart will have 3 halves, so at the end Bart has 3 times as many chestnuts as Adam has.

15 **(C) 3**
Any square has 4 vertices and any triangle has 3 vertices. 17 is neither a multiple of 4 nor a multiple of 3, so there is at least one square and one triangle on the table. Together, one square and one triangle have 7 vertices, so there are still 10 vertices not assigned to any particular figure. 10 is neither a multiple of 4 nor a multiple of 3, so there is yet another pair of a square and a triangle on the table having another 7 vertices altogether. There are still 10 − 7 = 3 unassigned vertices, so the rest of vertices belong to a triangle. Hence, there are 3 triangles on the table.

16 **(A) 2**

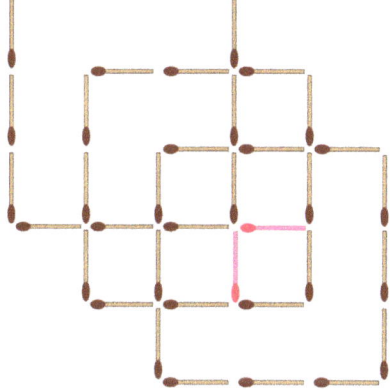

There are 8 squares visible in the original picture: 3 are large (3 × 3), 2 are medium (2 × 2) and 3 are small (1 × 1). By adding the two matches shown here in red, 3 more small squares will be added for a total of 11 squares altogether.

5 Point Solutions

17 (C)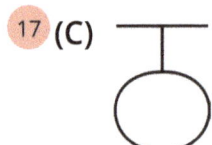

The images in the problem have the digit first facing the opposite the direction. Answer (C) shows the number 5 in this way.

18 (C)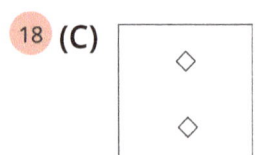

The top edge of the folded napkin is one line of symmetry and then, after the first unfolding, the edges on the left become one edge which is also a line of symmetry. In terms of the first line of symmetry, the paper cut-out looks like a diamond after the first unfolding and then, after unfolding along the second line of symmetry, we see two diamonds. After rotating the unfolded napkin by 90° we get the napkin shown in (C).

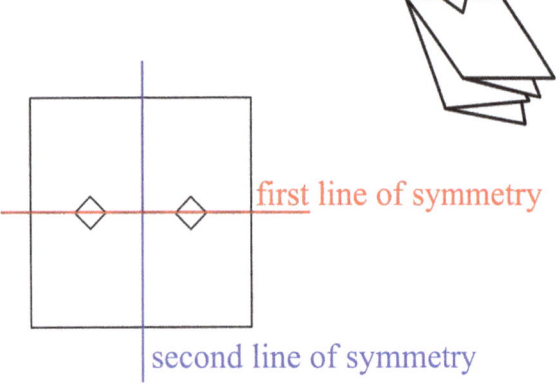

19 (D) 32

At the beginning, there were 12 – 8 = 4 more boys than girls. Since every week one more girl than boy was admitted, it will take 4 weeks for the numbers to be equal. After 4 weeks the number of girls will be 8 + 4 × 2 = 16 and the number of boys will be 12 + 4 × 1 = 16. The club will have 32 members then.

20 (A) 10

All the numbers that fulfill this requirement are (in the decreasing order): 400, 310, 301, 220, 211, 202, 130, 121, 112, and 103. There are 10 such numbers.

21 (A) 600 cm

The length of Anita's or Beth's towel is 720 cm ÷ 4 = 180 cm. Hence the side of the big square is 2 × 180 cm = 360 cm long. To get the width of one of the three smaller towels, divide 360 cm by 3, which is 120 cm. The length of one of these towels is 180 cm (half of the side of the big square). Thus, the perimeter of one rectangular towel is 2 × 120 cm + 2 × 180 cm = 600 cm.

22 (B) $5

Todd has 20 dollars and Will and Kevin together also have 20 dollars (20 + 20 = 40). Since Will has 10 dollars less than Kevin, he has 5 dollars and Kevin has 15 dollars (15 – 10 = 5; 5 + 15 = 20).

23 (D) 6

The table below records the number of candy bars in each basket (L = left, M = middle, R = right) after consecutive actions until the middle basket is empty.

L	M	R	M	L	M	R	M	L	M	R
10	10	10	9	9	8	9	7	8	6	8

M	L	M	R	M	L	M	R	M	L	M
5	7	4	7	3	6	2	6	1	5	0

The right basket contains 6 candy bars and the left basket contains 5 candy bars at the moment when the middle basket is empty, so 6 is the largest number of candy bars left.

24 (E) other number

Remember that the sum of the numbers on the opposite sides of a die is 7. This tells us which numbers are on each of the sides that we cannot see in the picture. So,
after the first move we will have
1 in front, 5 on top, and 3 to the right;
after the second move we will have
2 in front, 1 on top, and 3 to the right;
after the third move we will have
2 in front, 4 on top, and 1 to the right;
after the fourth move we will have
3 in front, 2 on top, and 1 to the right.
The final position is 2 on top which is "other number" from the options listed.

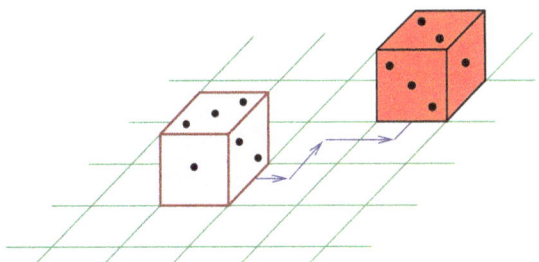

2003

2003

3 Point Solutions

1 (E) 12

Count the squares of the shaded part by using the grid to help.

---

2 (C) 4
To find the result of 0 + 1 + 2 + 3 + 4 − 3 − 2 − 1 − 0 quickly, notice that all the numbers that are added except for 4 are also subtracted.

---

3 (D) 160
Since each car has twice as many boxes as the car in front of it, multiply the number in each car by 2 to get the number of boxes in the next car.

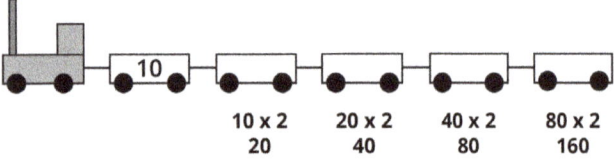

---

4 (A) blue

blue   green   red   yellow
 1   2   3   4
- 5 6 7 8 - 9 10 11 12 - 13 14 15 16 - 17

The pattern consists of four colors repeated in the same order. Because 17 is one more than 16, which is a multiple of 4, we can determine that the 17th color is the same as the first color in the sequence, which is blue.

---

5 (C) 50
Multiply the number of tables with the same number of chairs by the number of chairs next to the tables. 6 tables with 4 chairs each gives us 6 × 4 = 24 chairs. 4 tables with 2 chairs each gives us 4 × 2 = 8 chairs. 3 tables with 6 chairs each gives us 3 × 6 = 18 chairs. The total number of chairs in the teachers' lounge is 24 + 8 + 18 = 50 chairs.

---

6 (D)

♥ ♥ ♥
♥ ♥ ♥
♥ ♥ ♥
△ ◊ △

For every shape that is not a heart there have to be 3 hearts in order for the picture to have 3 times as many hearts as other shapes. Picture D has 9 hearts and 3 other shapes. 3 × 3 = 9.

---

7 (C) 10

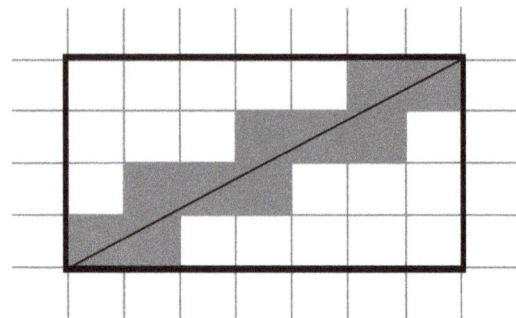

Draw a diagonal as shown and count the squares it passes through. The diagonal passes through 10 squares of the grid.

8. **(C) 21 grams**
There are 9 cubes. Since they are identical, each cube weighs the same amount. The weight of all the cubes is 189 grams or 189 g. The total weight divided by the number of cubes will equal the weight of one cube. 189 g divided by 9 equals to 21 g.

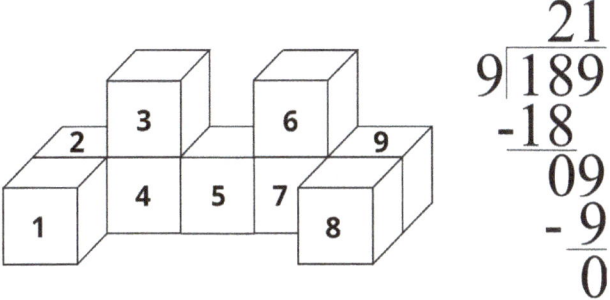

4 Point Solutions

9. **(C) 23**
Consecutive numbers are numbers in order such as 1, 2, 3, etc. Natural numbers are the normal counting numbers. Philip wrote down seven single-digit numbers from 3 to 9. He still needed to write 35 – 7 = 28 more digits, which means that he wrote fourteen more two-digit numbers, from 10 to 23, so 23 is the greatest number that Philip wrote down.

10. **(C) 11 hr 5 min**

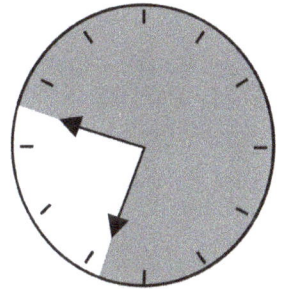

Anna slept from 9:30 p.m. to 6:45 a.m. as indicated by the shaded part of the clock, so she slept 9 hours and 15 minutes. Since Peter slept 1 hour and 50 minutes longer than Anna, he slept for 11 hours and 5 minutes. Remember that 15 minutes + 50 minutes is 65 minutes, which is 1 hour and 5 minutes.

11. **(D) 8**
We can describe the pattern as 8 pairs of black-white bars plus the last black bar, so there are 8 white bars in the pattern.

12. **(A) 6 m 78 cm**
55 dm = 55 × 10 cm = 550 cm and 50 mm = 5 × 10 mm = 5 cm, so Jumping Kangaroo's longest jump during the Olympics was (550 cm + 5 cm) + 123 cm = 678 cm, which is 6 m 78 cm.

13. **(C) 18203**
Perform the operations in reverse order. From the final result of 20003 subtract 2003, which equals 18000. To get the number which Paul chose at the beginning, add 203 to 18000. The result is 18203.

14. **(A) 24**
For the hour part, we need to find the greatest sum of the digits on the interval from 00 to 24. For the hours from 20 to 24, the largest sum of digits is 2 + 4 = 6. For the hours from 00 to 19 the largest sum is 1 + 9 = 10, so 10 is the largest sum for the hour part. For the minute part, we need to find the greatest sum of the digits on the interval from 00 to 59, which happens to be 59, with 5 + 9 = 14. Barbara will get the greatest sum of the digits on her electronic watch at 19:59 (which is 7:59 p.m.), and that sum is 1 + 9 + 5 + 9 = 24.

15. **(B) 24**
24 is the number that would double the current number of apples. So, the number of apples Mark actually did pick is also 24.

16 (C) 3 m

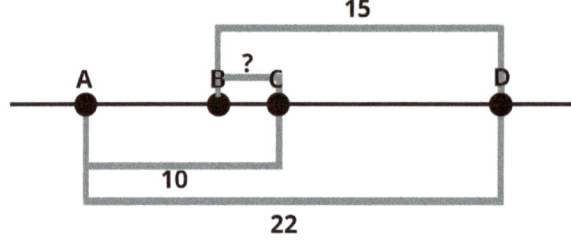

Since the distance from A to D is 22 m, and the distance from A to C is 10 m, then the distance from C to D is 22 m − 10 m = 12 m. Since the distance from B to D is 15 m, and the distance from C to D is 12 m, then the distance from B to C is 15 m − 12 m = 3 m.

5 Point Solutions

17 (D) 4
Three students do not have siblings, therefore there are 29 − 3 = 26 students who have siblings. The number of sisters and brothers the students have is 12 + 18 = 30 siblings. Because there are more siblings than students, there will be some students who have both a sister and a brother. There are 30 − 26 = 4 students with both a sister and a brother.

18 (E) 9
Once Daniel cuts a piece of paper into three parts, instead of having one piece he now has three. That is two more pieces than he started with, even though they are now smaller.
If we cut 1 piece of paper,
we have 11 + 2 pieces.
If we cut 2 pieces of paper,
we have 11 + 2 + 2 pieces.
If we cut 3 pieces of paper,
we have 11 + 2 + 2 + 2 pieces, and so on.

Thus, 11 + 2 × the number of all pieces cut is 29. Hence, the number of all pieces cut is half of 29 − 11 = 18, which is 9.

19 (E) 1
John bought 3 kinds of cookies: large, medium, and small, so he had to buy at least one of each and pay 4 + 2 + 1 = 7 dollars for these 3 cookies. After that he had 9 dollars left to buy 7 more cookies. 6 cookies of any kind cost at least 6 dollars, so the seventh cookie cannot be a large one since 6 dollars + 4 dollars is 10 dollars (only 9 dollars are left). Therefore, John bought only 1 large cookie. (If we solve the problem further, we can find that he bought 1 large cookie, 3 medium cookies, and 6 small cookies.)

20 (A) 12
Christopher built a rectangular prism with the dimensions of 5 × 4 × 4. To find the number of all the blue inner cubes, we need to remove all the red outer walls (or layers): front and back, top and bottom, left and right sides. Basically, we are reducing the dimensions of Christopher's rectangular prism by 2 units for each dimension. The resulting dimensions are 3 × 2 × 2, so the number of blue inner cubes is 12.

21 (B) 16
Jerry's purchase plans differ by 2 basketballs. To buy those two additional basketballs he needs to borrow 22 dollars and also use the 10 dollars he would have left over had he only bought 5 basketballs. The cost of buying the two extra basketballs is 22 + 10 = 32 dollars. Therefore, one basketball costs 16 dollars.

22 (D)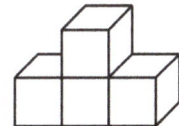

Notice that the bottom of the back wall is made of 3 white cubes and another white cube is visible above its center.

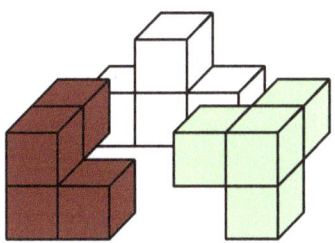

---

23 (A) 1 and 3

Draw in the lines of the grid over the shaded region. Piece 1 is the only one that can fill in the bottom of the shaded region; it is marked in blue here. Piece 3 needs to be rotated clockwise to fill in the remainder of the shaded region; it is marked in pink.

---

24 (B) **A bear is twice as expensive as a dog.** 3 dogs and 2 bears together cost as much as 1 dog and 3 bears since each collection costs as much as 4 kangaroos. Removing 1 dog and 2 bears from each collection results in 2 dogs being priced as much as 1 bear, so a bear is twice as expensive as a dog.

2005

2005

3 Point Solutions

1 (C) 500

$2005 - 205 = 1300 +$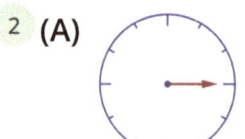

2005 − 205 = 1800
1300 + **500** = 1800

2 (A)

1 quarter of an hour = 15 minutes
1 hour = 4 quarters of an hour
17 quarters of an hour = 4 hours and 1 quarter of an hour = 4 hours and 15 minutes.

The time after 4 hours and 15 minutes past noon is 4:15. The minute hand must be one quarter of the way around the face of the clock.

3 (B) 3
Since Joan got $1 in change, she paid $10 − $1 = $9 for all the cookies.
Each cookie costs $3 each, so she bought 9 ÷ 3 = 3 cookies.

4 (B) 2
After the first whistle there were 4 monkeys in each of the 4 rows, so there were 4 × 4 = 16 monkeys at the circus. After the second whistle, the 16 monkeys formed 8 rows, so there were 2 monkeys in each row.

5 (B) 24
Eva's legs: 2
Parents' legs: 2 × 2 = 4
Brother's legs: 2
Dog's legs: 4
Cats' legs: 2 × 4 = 8
Parrots' legs: 2 × 2 = 4
Fish legs: 0
The sum of all the legs is
2 + 4 + 2 + 4 + 8 + 4 + 0 = 24.

6 (D) 60

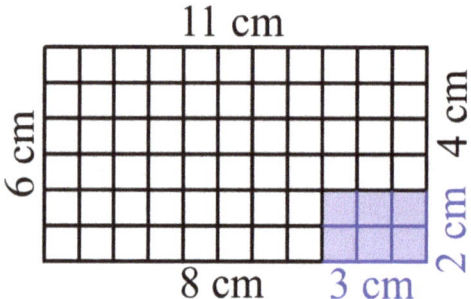

The lengths marked in centimeters (cm) also give us the number of rows and columns of pieces of chocolate. The whole bar had 11 × 6 = 66 pieces. The lengths of the sides of the part he ate are 11 − 8 = 3 and 6 − 4 = 2, so he ate 3 × 2 = 6 pieces. Starting with 66 pieces and eating 6 means that there are 66 − 6 = 60 pieces left.

7 (E) a truck that is 325 cm wide and weighs 4250 kg
Both the width and the weight of the trucks must be equal to or less to the numbers posted on the signs. The width of the truck has to be at most 325 cm AND at the same time the weight has to be at most 4300 kg. Of the trucks listed, only one that is 325 cm wide and weighs 4250 kg is allowed to cross the bridge.

8 (C) 5
The sum of 7 identical numbers must be a multiple of 7. The three-digit number is also a multiple of 10 because it ends in 0. Among the multiples of 10 from 300 to 390 only 350 is a multiple of 7, so the middle digit is 5. Indeed, 50 + 50 + 50 + 50 + 50 + 50 + 50 = 350.

4 Point Solutions

9 (C) 4
In such a family, there have to be at least two boys and two girls. If there was only one girl, she wouldn't have any sisters, and if there was only one boy, he wouldn't have any brothers. There are at least 4 children in this family.

10 (B) 3874
The number 4683 is not even. The number 4874 does not have all different digits. In the number 1246, the hundreds digit in not double the ones digit. Finally, the number 8462 does not have its tens digit greater than its thousands digit. Only the number 3874 satisfies all four conditions. It is even, all the digits are different, 8 is the double of 4, and 7 is greater than 3.

11 (B)
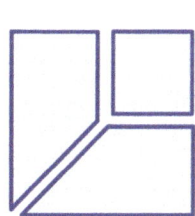

12 (C) 3
Two friends who weigh 80 kg each cannot take the elevator together, because 80 kg + 80 kg = 160 kg, which is more than 150 kg. Only the one friend who weighs 60 kg can take the elevator together with one of the other friends, because 60 kg + 80 kg = 140 kg. The other two friends who weigh 80 kg have to take the elevator separately. So, the four friends need to take at least 3 trips.

13 (D) 45
The difference between the amount of money that Sophia has, compared to Ala and Barb, is the same. Barb has $66 − $24 = $42 more than Ala, so Sophia has half of this difference ($21) more than Ala and $21 less than Barb. Sophia has $24 + $21 = $45.

14 (D) 1
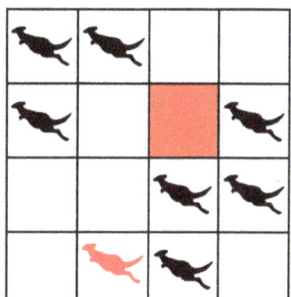

The kangaroo from the 3rd cell in the 2nd row needs to be moved to the 2nd cell in the 4th row.

15 (E) 8
If the sack didn't have a hole, Greg would have to make just 4 trips. Since he loses half the amount of sand from the sack, the number of his trips needs to be doubled to 8.

16 (D) after 10 days

In 4 days, Adam solved 4 × 5 = 20 problems. Solving 2 problems daily, Brad needed 10 days to solve 20 problems.

5 Point Solutions

17 (B) 3

There are now 15 − 9 = 6 more pieces of paper than before. Cutting one piece of paper into 3 pieces increases the number of all pieces of paper by 2. So, to make 6 new pieces, one needs to cut 3 pieces of paper.

18 (B) 3

Notice that because you're using whole matches, the length and the width need to be whole numbers. Because the perimeter is equal to two times the length plus two times the width, you can just consider what pair of numbers for the length and for the width would add up to 7, which is half of the whole perimeter.

There are 3 possibilities:
1. A rectangle with sides of 1 match × 6 matches
2. A rectangle with sides of 2 matches × 5 matches
3. A rectangle with sides of 3 matches × 4 matches

19 (A) 1 dm

The difference between the outside and the inside perimeters consists of 8 segments (2 at each corner). The length of each one is the same as the width of the frame. Thus, 8 × the width of the frame is 8 decimeters, so the width of the frame is 1 dm.

20 (D) 8

The following steps have to be taken to collect 50 coins:

- Step 1: The trunk needs to be open: **1 lock**.

- Step 2: 1st chest needs to be open: **1 lock**.

- Step 3: Three boxes in the 1st chest: **3 locks**. At this point, 30 coins can be taken out.

- Step 4: 2nd chest needs to be open: **1 lock**.

- Step 5: Two of the boxes in the 2nd chest: **2 locks**. 20 more coins can be taken out.

At least 1 + 1 + 3 + 1 + 2 = 8 locks must be open to take out 50 coins.

21. (C) 24 m

The lengths are indicated by red arrows and the width is indicated by the blue arrow in the picture. Two lengths cover the distance of 16 m, so the length of one flowerbed is 8 m.
The width of one flowerbed is
20 m – 8 m – 8 m = 4 m.
The perimeter of each flowerbed is
(2 × 8 m) + (2 × 4 m) = 24 m.

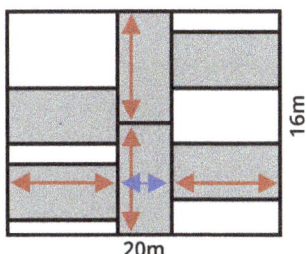

22. (E) 1009

Every two-digit number other than 10 added to 989 gives a four-digit number, so 10 is the only option for the two-digit number. The three-digit number must be 989 + 10, which is 999.
The sum of those numbers is
999 + 10 = 1009.

23. (B) 3

None of the cards is in its correct positions, so each of the cards must be moved. We simply switch two cards in one move. Only one move is needed to place 3 and 4 in their correct position. Just one more additional move will not fix the position of the other three cards, so at least two more moves are needed. Three moves is the least number of moves needed to place all the five cards in the correct order. One of the switching options is shown below.

1st move: Switch the cards 3 and 4
2nd move: Switch the cards 1 and 5
3rd move: Switch the cards 2 and 5.

24. (E)

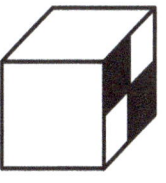

Notice that the four small squares will make a checkered pattern on the side opposite the one that is completely black. This excludes the first four options. (E) can happen, since the face opposite the checkered face, which we can't see, could be black.

2007

3 Point Solutions

1 (C) 2, 3, 5

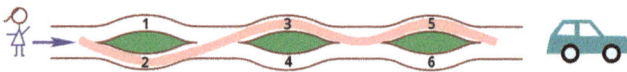

- Anna can't pick up 1 and 2 together, so (A) and (E) are eliminated.
- She can't pick up 3 and 4 together, so (B) is eliminated.
- She can't pick up 5 and 6 together, so (D) is eliminated.
- Along the way Anna could pick up 2, 3 and 5.

2 (A) 1

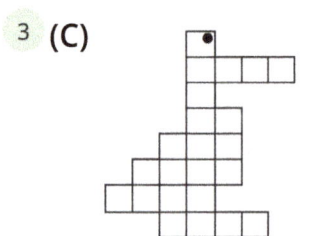

Only one letter is repeated; it is R.

3 (C)

Let's focus on rows 2, 5, and 6. In the other rows, the numbers of squares are identical.

Following is a count of squares in those three rows.

Row 2:	3	3	4	3	4
Row 5:	3	3	3	4	3
Row 6:	3	4	4	3	3
Sums:	9	10	11	10	10

The figure (C) consists of the largest number of squares.

4 (C) 3

Five notebooks cost 5 × 80 cents = 400 cents, or $4. Helen has $5, so after buying five notebooks she has $5 − $4 = $1 left to spend on pencils. She can buy at most 3 pencils for 30 cents each since 3 × 30 cents = 90 cents.

5 (C) 64

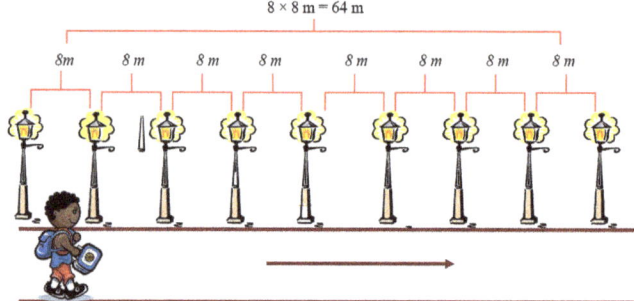

Between the first and the ninth lamppost, there are eight equal distances of 8 m each.

6 (E) 6

Here are all the possible codes:
1 3 5
1 5 3
3 1 5
3 5 1
5 1 3
5 3 1

7 (C) 48

Remember that you need to do multiplication before addition:

4 × 4 + 4 + 4 + 4 + 4 + 4 × 4 = 16 + 4 + 4 + 4 + 4 + 16 = 16 + 16 + 16 = 48

You might also notice that there are four 4's in the middle, which can make up another 4 × 4.

4 × 4 + 4 + 4 + 4 + 4 + 4 × 4 =
4 × 4 +   4 × 4   + 4 × 4 = 3 × (4 × 4) = 3 × 16 = 48

8 (B)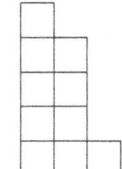

The given figure contains 8 unit squares. (A) contains 8 unit squares, (B) contains 10 unit squares, (C) contains 9 unit squares, (D) contains 6 unit squares, and (E) contains 9 unit squares.

Thus, the figure which is given together with any other figure contain 14 to 17 unit squares. Any rectangle containing the original figure must have one side at least 5 squares long and the other side at least 3 squares long, so such a rectangle must contain at least 15 = 5 × 3 unit squares. This eliminates (D) as an option since 8 + 6 = 14 is less than 15.

The given figure and (A) together have 16 unit squares. There are several ways to make a rectangle which contains 16 square units: 16 = 16 × 1, 16 = 8 × 2, and 16 = 4 × 4. In the first case 1 is less than 3, in the second case 2 is less than 3, and in the third case both sides are less than 5, so the original and (A) cannot form a rectangle. The same is true for (C) and (E) since 8 + 9 = 17 and any rectangle containing 17 unit squares must have dimensions 17 × 1. Of course, 1 is less than 3 which eliminates (C) and (E) as valid options.

(B) is the only option left and we can bring together (B) and the given figure to see a rectangle. We need to rotate (B) counterclockwise as shown below in blue, and then put the two figures together.

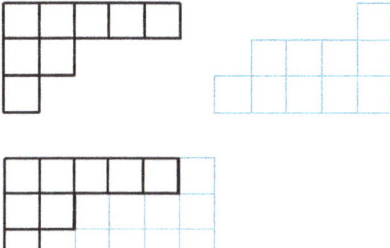

4 Point Solutions

9 (C) 5

To find the number in the shaded cloud, start with the last cloud and reverse the operations.

10 (C) 3
Each of the digits 1, 2, and 3 appears in each row and in each column once and only once, so 3 must be the last entry in the first column. The second entry in the last row is also the third entry in the second column (marked in gray), so that entry is neither 3 nor 1. It must be 2, so in the square marked with the question mark Harry can only write 3. After that there is only one way to fill in the big square, which is shown below.

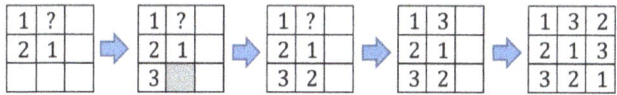

11 (A) 2016
The sum of the digits of 2007 is equal to 9. The next two numbers are 2008 and 2009. For each of those numbers the sum of the digits is greater than 9. After that the tens digit becomes 1, so the next number with the sum of digits equal to 9 is 2016, which is the answer.

12 (C) 17
Annette needs 27 blocks to fill up the whole aquarium since
Length × Width × Depth = 3 × 3 × 3 = 27. She already put in 10 blocks, the 6 we can see, and 3 more in the bottom layer and 1 in the middle layer that are hidden behind them. 27 − 10 = 17, so she still needs to add 17 cube blocks.

13 (A) January 2nd, 2003

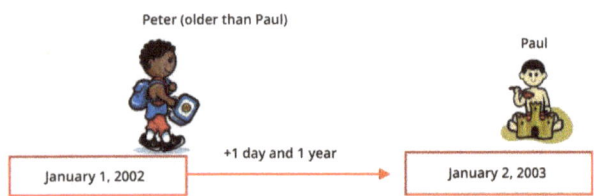

Peter is older than Paul, so he was born first. Add 1 year and 1 day to Peter's birthdate to get Paul's.

14 (B) 60 m
400 × 15 cm = 6000 cm, and we know that 1 m = 100 cm, so 6000 cm = 60 m.

15 (B) 5
Working backwards, 72 − 19 = 53, which gives us the two-digit number. David wrote the digit on the left (the tens digit) first, so the first digit he wrote was 5.

16 (A) 4 hr 55 min
Rearranging the digital time only using the digits 0, 0, 2, and 7, we could get 02:70, which is not a valid time since the minute part can only be a number from 00 to 59. We could also get 7:02 and 7:20, the earlier of which is 7:02. The difference in time between 07:02 and 02:07 is the difference between 6 hours + 62 minutes and 2 hours + 7 minutes. This difference is 4 hours + 55 minutes, so (A) is the answer.

5 Point Solutions

17 (E) 12

At each edge of the large cube, there is exactly one small cube (the middle one) with exactly two gray sides. Every cube has 12 edges (4 parallel edges in each direction), so there are 12 small cubes that have exactly two gray sides.

18 (B) after 110 km
15951 is the highest palindromic number among numbers of the form 15☐51, so the next palindromic number has the form 16☐61. The smallest among them is 16061 and 16061−15951 = 110, so after 110 km the odometer will show a palindromic number again.

19 (C) 65
When we move from one big square to the next one, we add two more rows and two more columns, so the fourth square must have 9 rows and 9 columns. So, the fourth square contains 9 × 9 = 81 small squares. Remove all the small white squares from all big squares. The first big square is left with 1 row and 1 column of gray squares, the second big square is left with 2 rows and 2 columns of gray squares, and the third big square is left with 3 rows and 3 columns of gray squares. Following this pattern, the fourth big square is left with 4 rows and 4 columns of gray squares. Thus, the fourth big square contains 4 × 4 = 16 gray squares. Therefore, the number of white squares in the fourth big square is 81−16 = 65.

20 (A) 1st

Adam   Celina   Bob   Daniel   Eve

Daniel is ahead of Celina and nobody is standing between Bob and Daniel, so Celina is behind Bob and Daniel. Adam is behind Celina, so the four of them are standing in the following order (counting from the farthest away from the cashier): Adam, Celina, Bob, and Daniel. Daniel is not the first in the line, so Eve must be the first one.

21 (A) 48 cm

The perimeter of the rectangle is 2 × (15 + 9) = 2 × 24 = 48 cm. The perimeter of the rectangle and the perimeter of the polygon obtained by cutting off four corners of the rectangle share the four black segments. The eight red segments of the perimeter of the rectangle match the eight blue segments of the perimeter of the polygon, so the perimeter of the polygon is the same as the perimeter of the rectangle, which is 48 cm.

22 (B) 14

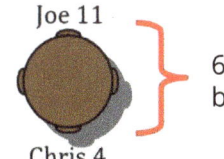

6 chairs on each side between Joe and Chris.

There are 6 chairs (numbered from 5 to 10) on one side of the table between Chris and Joe and the same number of chairs on the other side. 2 × 6 chairs + Chris's chair + Joe's chair = 12 + 1 + 1 = 14 chairs.

23 (D) 192

9 digits are used to write one-digit numbers (1, 2, ... , 9). There are 90 two-digit numbers (from 10 to 99), so 2 × 90 = 180 digits are used to write all two-digit numbers. Finally, 3 digits are used to write the number 100. 9 + 180 + 3 = 192 digits.

24 (E)

Each of the pictures (A), (B), (C), (D) can represent this unfolded sheet of paper.
The sheet of paper is folded twice. Below is another drawing of the folded sheet. Let's make the upper left-hand corner the point where the two fold lines cross. Let's flatten the folded paper to form the smaller square and number the vertices clockwise starting with 1 at the upper left-hand corner. The paper was rotated, so that we do not know which corner was cut. We need to think through the problem taking one corner at a time.

By cutting off corner 1 and unfolding the paper we will see a hole at the center of the paper since the two folding lines intersect at the vertex 1. This is the shape shown by (D).

Vertex 2 is at the two endpoints of the horizontal folding line. If we cut off corner 2 and unfold the paper, then we see shape (C).

Vertex 3 represents the four vertices of the original square sheet of paper. If we cut off corner 3 and unfold the paper, then we see the four vertices of the original square sheet of paper cut off. This is the shape shown by (A).

Vertex 4 is at the two endpoints of the vertical folding line. If we cut off corner 4 and unfold the paper, then we see shape (B).
In conclusion, we can get all four shapes.

2009

2009

3 Point Solutions

1 **(E) 15**
There are 3 wooden cubes stacked on top of each other in each of the 5 columns, so 3 × 5 = 15 wooden cubes were used.

2 **(C) 2009**
Perform the multiplication first, to get 200 × 9 + 200 + 9 = 1800 + 200 + 9 = 2009.

3 **(B) In the circle and in the square, but not in the triangle.**

The kangaroo is in the region where the circle and square overlap.

4 **(A) 6**
Each of the brothers has the same girl as a sister. So, there is only one girl in the family. 5 boys + 1 girl = 6 children in this family.

5 **(B) 6**

To change the hundreds digit from 9 to 8, only one little square needs to change from white to gray. To change the tens digit from 3 to 0, two little squares need to change from white to gray and one little square from gray to white. To change the ones digit from 0 to 6, one little square needs to change from white to gray and another one from gray to white. There are 6 little squares that need to change color.

6 **(B) 6**
Carl ate half of 16 oranges, so he ate 8 oranges. From the remaining 8 oranges, Eva ate 2, so Sophie ate 6 oranges.

7 **(C) 46 m**
The black zigzag line is made of 5 longer segments and 4 shorter segments.

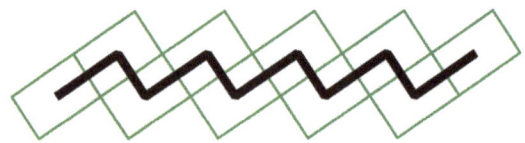

Each longer black segment connects midpoints of two adjacent tiles in the direction of longer sides, so its length is exactly the length of one tile, which is 6 meters. Each shorter black segment connects midpoints of two adjacent tiles in the direction of shorter sides, so its length is exactly the width of one tile, which is 4 meters. The length of the whole black path is 5 × 6 m + 4 × 4 m = 30 m + 16 m = 46 meters.

8 (D) at 6:53 p.m.
With the commercial breaks, it took 90 + 8 + 5 = 103 minutes for the movie to finish, which is 1 hour and 43 minutes. The movie started at 5:10 p.m., so 1 hour and 43 minutes later means the movie ended at 6:53 p.m.

4 Point Solutions

9 (C) 87 kilograms
The weight difference between the heavier gray kangaroo and the lighter red kangaroo is 35 kilograms. The combined weight of both the gray and the red kangaroos is 139 kilograms. Adding their difference of 35 kilograms to their combined weight is equivalent to twice the weight of the heavier gray kangaroo. 139 kilograms + 35 kilograms = 174 kilograms, so half of it, which is 87 kilograms, is the weight of the gray kangaroo. The lighter red kangaroo weighs 35 kilograms less, which is 52 kilograms.

10 (D) 40

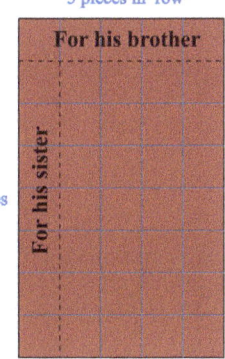

Since Zach first broke off one row of 5 pieces for his brother, the chocolate bar was 5 pieces wide.
After that Zach broke off one row of 7 pieces for his sister, but at this point 1 row of pieces was already gone. So, the whole chocolate bar had 8 rows going across. There were 5 × 8 = 40 pieces in the whole chocolate bar.

11 (A) 6
There were 25 − 19 = 6 more boys than girls in a dance group to start. Each week the difference between the number of boys and girls decreased by one, since for every 2 boys there are 3 girls joining. The number of boys and girls will be the same after 6 weeks.

12 (B) 90
Each cow has 4 legs and each chicken has only 2 legs. For each cow there need to be two chickens for the number of legs to be equal. Since there are 30 cows, there are 2 × 30 = 60 chickens. The farmer has 30 + 60 = 90 animals altogether.

13 (B) 6 cm
One side of the rectangle is 8 cm and the other side is half of 8 cm, which is 4 cm. The perimeter of the rectangle, or the distance around it, is 2 × 8 cm + 2 × 4 cm = 16 cm + 8 cm = 24 cm. The perimeter of the square is the same as that of the rectangle, and the square has four sides of equal length, so length of one side of the square is 24 cm ÷ 4 = 6 cm.

14 (D) 3
If Magda rolled the die four times and got 6 on each roll she would have obtained 24 points. That would be 1 point more than the 23 points she got, so one of the rolls of the die had to show 5 dots and the other three rolls had to show a 6. There is no other way to get 23 with four die rolls.

15 (B) 2

We need to find three different whole numbers that add up to 7. The middle number will be how many nuts Tola, who did not find the smallest or the greatest number of nuts, found. Trying with the smallest numbers possible, we quickly see that the only numbers that work are 1, 2, and 4. The middle number is 2.

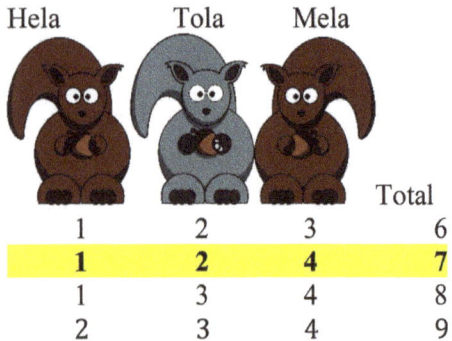

Hela	Tola	Mela	Total
1	2	3	6
1	**2**	**4**	**7**
1	3	4	8
2	3	4	9

---

16 (A) 6

Paul had 27 scouts on one side and 13 scouts on the other side of him, so there were 27 + 1 + 13 = 41 scouts standing in a single row. Peter was standing exactly in the middle of the row, so he had 20 scouts on each side of him. Among the 20 scouts on one side of Peter there are 13 scouts + Paul + the scouts between them, so the number of scouts between Peter and Paul is 6.

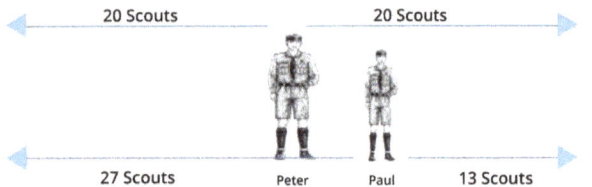

5 Point Solutions

17 (E)

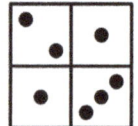

All the possible arrangements of the two dominoes in vertical and horizontal positions are shown below.

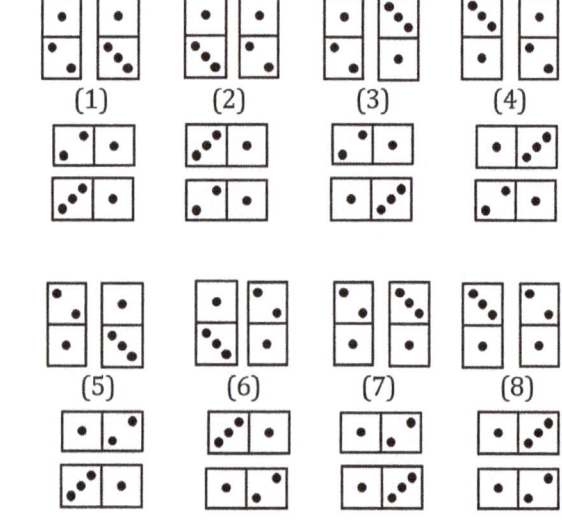

Both 2 and 3 dots on the dominoes are always slanted in the same direction, so we cannot get figure (E) as a result.

(A) is a vertical (1), (B) is a vertical (5), (C) is a horizontal (7), and (D) is a horizontal (3).

18 (D)

In the place of each unknown digit, marked by ✻, we can put any digits from 0 to 9. For (A), the sum of the first, third, and fifth digits is 8 + ✻ + 6 = 14 + ✻, and it can be any number from 14 to 23, depending on which digit we chose; adding 0 gives the lowest value and 9 gives the highest. The sum of the second, fourth, and sixth digits is 1 + ✻ + 1 = 2 + ✻, and it can be any number from 2 to 11, so the first sum can never be equal to the second sum.

For (B), the first sum is 7 + 7 + 7 = 21 and the second sum is ✻ + 2 + ✻ = 2 + ✻ + ✻. The second sum can be any number from 2 to 2 + 9 + 9 = 20, but it is never 21.

For (C), the first sum is 4 + 4 + 4 = 12, and the second sum is ✻ + 1 + 1 = 2 + ✻. The second sum can be any number from 2 to 11, but it is never 12.

For (D), the first sum is 1 + ✻ + ✻, and it can be any number from 1 to 1 + 9 + 9 = 19. The second sum is 2 + 9 + 8 = 19. The two sums can be equal.

This happens when the code is

| 1 | 2 | 9 | 9 | 9 | 8 |

For (E), the first sum is 1 + 1 + 2 = 4, and the second sum is 8 + ✻ + ✻. The second sum can be any number from 8 to 8 + 9 + 9 = 26, but it is never 4.

Only the sums in (D) can be equal and only when the code is

| 1 | 2 | 9 | 9 | 9 | 8 |

19 (B) 24

The number of eggs Ms. Florentina sold on Tuesday is the difference between the number of eggs she sold on Thursday and the number of eggs she sold on Wednesday, so Ms. Florentina sold 96 − 60 = 36 eggs on Tuesday. The number of eggs she sold on Monday is the difference between the number of eggs she sold on Wednesday and the number she sold on Tuesday, so she sold 60 − 36 = 24 eggs on Monday.

20 (D) 4

Here are the possible ways that Kaya could have gone:

red → blue → white → yellow
(but she cannot go red → blue → yellow → white)

red → yellow → blue → white
(but she cannot go red → yellow → white → blue)

red → white → blue → yellow

red → white → yellow → blue.

21 (A) 5:00 p.m.

It is 12 hours from 6:15 a.m. to 6:15 p.m., and then another 1 hour and 15 minutes to 7:30 p.m., so Jasper was gone 13 hours and 15 minutes. The clock had been running backwards for the same amount of time, so it went back 12 hours to 6:15 p.m., and then back another 1 hour and 15 minutes to 5:00 p.m.

22 (B) 2

In the original table the sums in the rows are 10, 22, and 10. None of them is a multiple of 3. After any first move there is at least one unchanged row, so the sum of the number in it still is not a multiple of 3. Thus, at least two moves are needed to get the table we want. There are many ways of fixing the table in two moves. Here is a one way: in the first move switch 5 and 7, and in the second move switch 8 and 1 from the third row.

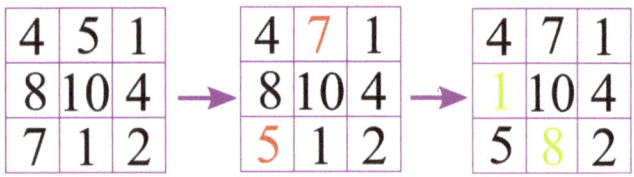

23 (E)

First, let's convert ♥ ♠ ♠ ♠ ♥ ♥ into words. The pattern is: right, left, left, left, right, right. All segments are either vertical or horizontal, so the first segment can be drawn vertically up, vertically down, horizontally right, or horizontally left. After that, the pattern of turns tells us what the figure looks like. In the pictures below, the black arrow shows the first segment, orange arrows show segments drawn after turning to the right, and blue arrows show segments drawn after turning to the left. In each case, we are following the instructions "right, left, left, left, right, right."

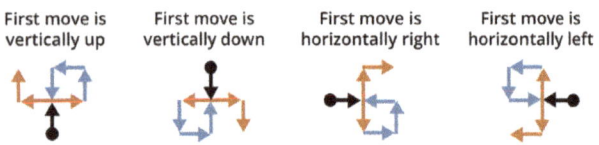

The first figure is exactly the figure in answer (E). The other figures, after rotating, also match the figure (E), so Agnes could have drawn only (E).

24 (B) 5

Among all shoes bought by a group of friends, only the smallest right shoe and only the largest left shoe will be left over if only two shoes are left. All pairs of shoes bought by a group of friends have sizes between 36 and 45. 36 is an even number and 45 is an odd number, so we cannot go from size 36 to 45 by adding only 2 shoe sizes, which would happen if only men were in the group. At least once we have to add 1 shoe size. It indicates that at least one person among the group of friends is a woman. Any two additions of 1 shoe size can be replaced by one addition of 2 shoe sizes, or any two women's shoe sizes can be replaced by one men's shoe size. For the smallest number of people there must be only one woman among the group of friends. The options for the shoe sizes are:

36→37→39→41→43→45,
36→38→39→41→43→45,
36→38→40→41→43→45,
36→38→40→42→43→45
and 36→38→40→42→44 → 45.

In each case there were 6 pairs of shoes shared among 5 friends.

2011

2011

3 Point Solutions

1 (A) 20 + 11

(A) 20 + 11 = 31
(B) 20 − 11 = 9
(C) 20 + 1 + 1 = 22
(D) 20 − 1 − 1 = 18
(E) 2 + 0 + 1 + 1 = 4
31 is the greatest number.

2 (C) Wednesday
The word KANGAROO has 8 letters. Michael will paint the letters on the following days: K on Wednesday, A on Thursday, N on Friday, G on Saturday, A on Sunday, R on Monday, O on Tuesday, and finally the last O on Wednesday.

3 (C)

The left side of the scale weighs 26 + 12 + 8 = 46 kg, and the right side of the scale is currently loaded with 20 + 17 = 37 kg. To balance the scale, we need to add 9 (because 46 − 37 = 9 kg) to the box on the right side of the scale.

4 (E) six
Paul got up 2½ hours ago and still has 3½ hours before the train leaves, so he got up 2½ + 3½ = 6 hours before it leaves.

5 (B)

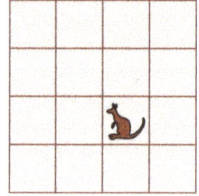

After the first four moves (one square to the right, one square up, one square to the left, one square down), the toy ended up in exactly the same place as at the beginning. The child then moved it one more square to the right, ending one square to the right from the original location.

6 (A) 3 dollars and 30 cents
3 scoops of ice cream cost $4.50, so 1 scoop costs $4.50 ÷ 3 = $1.50. 2 cookies cost $3.60, so 1 cookie costs $3.60 ÷ 2 = $1.80. Ala paid for one scoop and one cookie, so she paid $1.50 + $1.80 = $3.30.

7 (B)

There are only two gray figures, a triangle and a rectangle. Susan described the one which is not a rectangle. Therefore, she was talking about the gray triangle.

8 (D) 30
Between 7:45 and 10:45 the clock will strike on the hour at 8:00 (8 times), at 9:00 (9 times), and at 10:00 (10 times). Also, it will strike at half past the hour at 8:30 (1 time), at 9:30 (1 time), and at 10:30 (1 time). The clock will strike 8 + 9 + 10 + 1 + 1 + 1 = 30 times.

4 Point Solutions

9 (C)

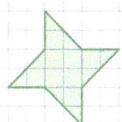

For each figure, count the number of "full" squares and "half-squares." The area of 2 half-squares is equal to 1 full square.

(A) 8 full squares + 4 half-squares =
8 squares + 2 squares = 10 squares
(B) 8 full squares + 6 half-squares =
8 squares + 3 squares = 11 squares
**(C) 8 full squares + 8 half-squares =
8 squares + 4 squares = 12 squares**
(D) 8 full squares + 2 half-squares =
8 squares + 1 square = 9 squares
(E) 8 full squares + 2 half-squares =
8 squares + 1 square = 9 squares

Figure (C) has the greatest area.

10 (B) 6
To find the least number of boxes, the larger boxes need to be filled first. 12 eggs fit in a large box, so 5 large boxes will hold 5 × 12 = 60 eggs. The other 6 eggs will fit in 1 small box. The least number of boxes the farmer needs to store 66 eggs is 6.

11 (B) 12

There are 8 cats, 6 dogs, and 3 fish in the picture. Two students have a dog and a fish, three students have a cat and a dog, so these five students together have 3 cats, 5 dogs, and 2 fish. The remaining 5 cats, 1 dog, and 1 fish belong one each to 7 students. There are 5 students with multiple pets and 7 students with one pet each, so there are 12 students in the class.

12 (C) 21

Each cake was divided into 4 × 3 = 12 pieces. There were two identical cakes, so there were 2 × 12 = 24 pieces altogether. Each person got one piece and three pieces were left over, which means that there were 24 − 3 = 21 people at the party.

13 (E) E

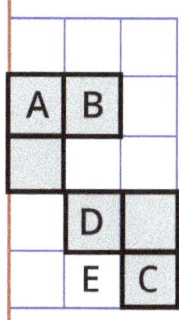

After folding the paper, the right side of the picture will look as shown.

14 (A) 60 cents
If all 13 coins are 10 cent coins, John has 130 cents. If all 13 coins are 5 cent coins, John has 65 cents, so he cannot have less money than that, such as 60 cents, in his pocket.
If John has twelve 10¢ coins and one 5¢ coin, the total value is 125 cents.
If he has ten 10¢ coins and three 5¢ coins, the total value is 115 cents.
If he has three 10¢ coins and ten 5¢ coins, the total value is 80 cents.
If he has one 10¢ coins and twelve 5¢ coins, the total value is 70 cents.

15 (D) 5
The possible results on a six-sided die are 1, 2, 3, 4, 5, and 6. If Ari got a result four times as great as Chuck, the only possibility is that Ari got 4 and Chuck got 1. If Darius got a result twice as great as Jack and three times as great as Mark, then it must be that Darius got 6, Jack got 3 and Mark got 2. Tom rolled to only number not mentioned yet, 5.

---

16 (C) 11

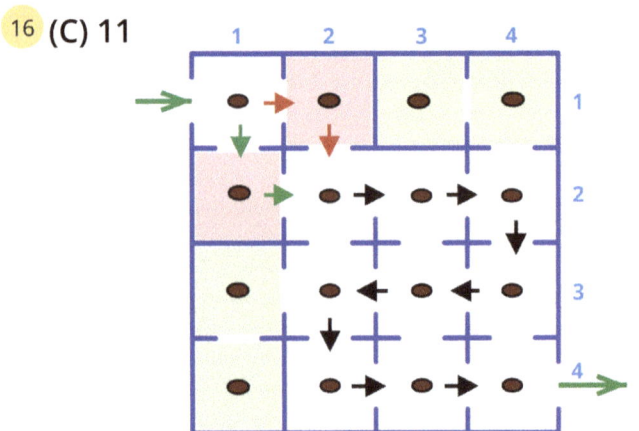

The squirrel will miss either the nut in row 1 column 2, or row 2 column 1 (shaded in pink). It will also not be able to get to the nuts in row 1 column 3, row 1 column 4, row 3 column 1, and row 4 column 1 (shaded in green). Since there are 16 nuts total and it will not be able to get at least 5 of them, the most nuts that the squirrel can get is 16 − 5 = 11. In the first two steps the squirrel can gather 2 nuts and after that the black arrows show one of the two possible paths for gathering 9 more nuts.

5 Point Solutions

17 (A) 7
If Mrs. Smith had answered exactly half of the questions correctly, her score would stay at 10 points since her wins and losses would cancel each other out. She got 14 points, so she answered more than half of the questions correctly. The incorrect answers were canceled out by some of the correct answers. In addition to the group of the incorrect and the correct answers which cancel them out, there are 4 more correct answers, because Mrs. Smith ended with 4 more points than she started with. Thus, the incorrect and the correct answers that cancel them out make up 10 − 4 = 6 answers. Half of these answers were incorrect, so Mrs. Smith answered 3 questions incorrectly, which means that she gave 10 − 3 = 7 correct answers.

---

18 (C) Dasha, Sasha, Pasha, Masha
Take a look at where the girls ended up and work backwards. The order at the end was Masha, Sasha, Dasha, Pasha. There were two changes of places. Since we're working backwards, first we need to look at the last change. Dasha changed places with Pasha, so the order before the initial switch was Masha, Sasha, Pasha, Dasha. In the first change of places, Masha changed places with Dasha, so the original order was Dasha, Sasha, Pasha, Masha.

---

19 (C) 10
From the diagram, we know that $AF = 34$ and $AD = 21$, so $DF = 34 − 21 = 13$. We also know that $CF = 23$ and since $DF = 13$, $CD = 23 − 13 = 10$. So, the distance between towns C and D is 10 miles.

20 (C)

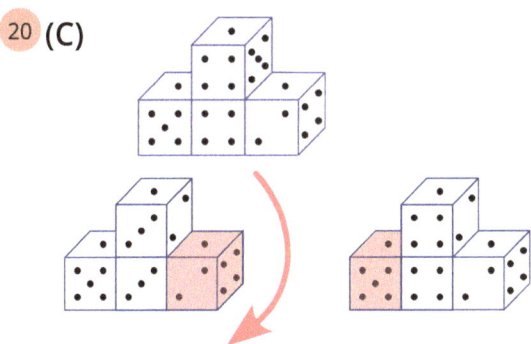

To see what the structure looks like from behind, we rotate it and focus on the colored die shown on the right. It will now be on the left with 1 dot still on the top, the initial front face (2 dots) in the back and the initial back face (5 dots) now in the front. The initial face with 4 dots will now be on the left side of the structure (not visible). The top die after the rotation will show 3 dots as the new front face and 5 dots as the new right face, thus showing structure (C).

---

21 (E) 12

A card that has a 6 can also be a 9, and a 9 can be a 6. Count all three-digit numbers that have two 6's and one 8, two 9's and one 8, or one 6, one 9, and one 8. These numbers are 668, 686, 866; 899, 989, 998; 689, 698, 869, 896, 968, and 986.

---

22 (D)

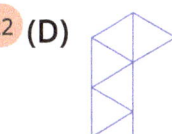

Figure (D) consists of 6 small triangles and the ornament consists of 18 small triangles, so we would need 3 pieces of (D) to build the ornament without any overlap. There are 6 positions of figure (D) that fit into the ornament. 3 of them can be put into the ornament in two ways which gives 9 configurations shown below.

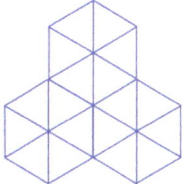

We cannot add two more pieces like that without overlap, so we cannot build the ornament using (D). The ornament can be built from each of the other shapes as shown below.

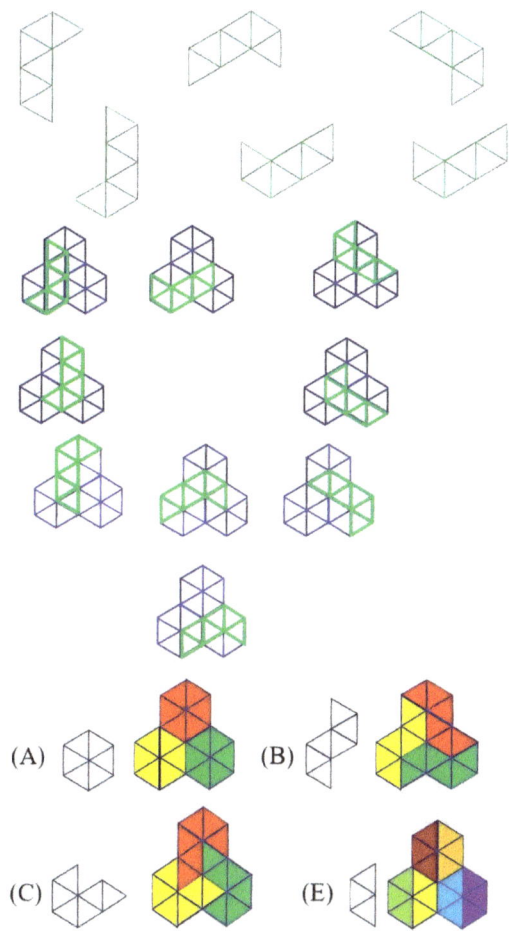

23 (A) 56

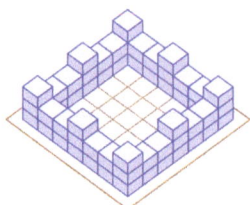

The castle has 8 taller columns made up of 3 cubes each (8 × 3 cubes = 24 cubes) and 16 columns made up of 2 cubes each (16 × 2 cubes = 32 cubes). 24 + 32 = 56 cubes were used to build the castle.

24 (E) 16

8 cannot be placed on the edge with 6 since 8 + 6 = 14 is greater than 13.
8 cannot be placed on the edge with 7 since 8 + 7 = 15 is greater than 13.
If 8 is placed in the white circle opposite 6, then the top edge contains only numbers less than 6, so its sum would not exceed 5 + 4 + 3 = 12, which is less than 13. If 8 is placed in the white circle opposite 7, then the left edge contains only numbers less than 6, so its sum would be less than 13. Therefore, 8 must be placed in the upper left corner.

5 cannot be placed on any edge containing 8 since 5 + 8 + a third number is always greater than 13, so the only place for 5 is the lower right corner. After that the other numbers are easy to calculate and place in the proper circles.

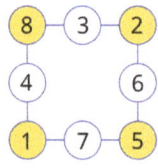

The sum of the numbers in the circles shaded yellow is 8 + 2 + 5 + 1 = 16.

2013

2013

3 Point Solutions

1 (D)

Figure (A) shows 3 black kangaroos and 3 white kangaroos.
Figure (B) shows 4 black kangaroos and 4 white kangaroos.
Figure (C) shows 4 black kangaroos and 4 white kangaroos.
Figure (D) shows 5 black kangaroos and only 4 white kangaroos.
Figure (E) shows 5 black kangaroos and 5 white kangaroos.

Figure (D) is the only figure with more black kangaroos than white kangaroos.

2 (D) 7

The sum of the two digits needs to end in 4. The options are 2 + 2 = 4 and 7 + 7 = 14. However, using 2 does not makes the calculation correct: 42 + 52 = 94, not 104. 47 + 57 = 104, so 7 is the digit under the stickers.

3 (E)

The pattern is such that after a certain number of black circles, there is the same number of gray circles, so the last 4 circles should all be shaded gray.

4 (B) 10

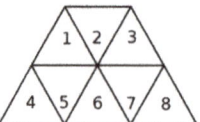 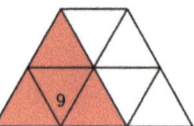 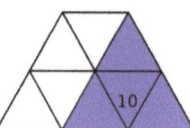

There are ten triangles in total, eight small ones numbered 1 to 8, and two big ones numbered 9 (red) and 10 (purple).

5 (C) 16

USA won a total of 46 + 29 + 29 = 104 medals, while China won a total of 38 + 27 + 23 = 88 medals, which gives a difference of 104 − 88 = 16 medals.

6 (D) 5

36 is divisible by 2, 3, 4, and 6 from the numbers listed, so Daniel can divide all the candy evenly among those numbers of friends. 36 is not divisible by 5. If Daniel tried to divide the candy equally among five friends, he would have to give each friend seven pieces, and have one piece left over.

7 (B) 30

Since each sandwich needs two slices of bread, Vero's mom can make 12 sandwiches from one package of bread. She can make 12 + 12 + 6 = 30 sandwiches from two and a half packages.

8 (E) Eddie

One of the digits of 325, the digit "2," is not odd, so Eddie was wrong.

4 Point Solutions

9 (B)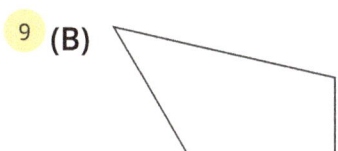

Extend the broken mirror to its original shape to see the missing piece shown in light blue. Rotate the missing piece around to match it with the piece (B).

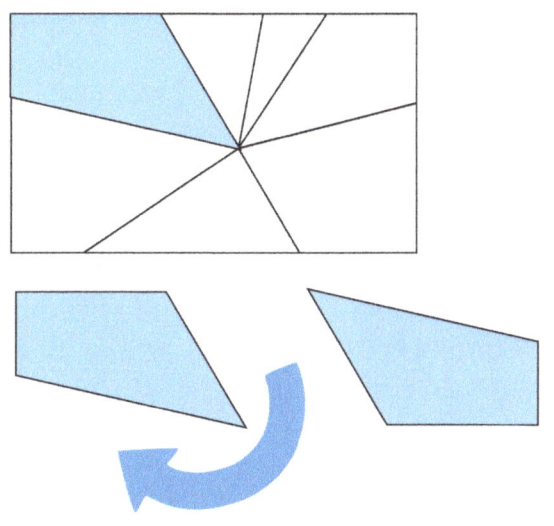

10 (D) 23 cm
Since three lies add 3 × 6 cm = 18 cm to the length of his nose, and two true statements shorten it by 2 × 2 cm = 4 cm, the final length of Pinocchio's nose is 9 cm + 18 cm − 4 cm = 23 cm.

11 (D) 5
Pedro cannot buy just four boxes of oranges since buying four boxes of the largest size would give him only 40 oranges. However, if Pedro buys three boxes of 10 and two boxes of 9 oranges, then he will have exactly 3 × 10 + 2 × 9 = 48 oranges in five boxes.

12 (A)

Ann's turns have been marked in the figure. After all of the shown turns, she is walking towards the train.

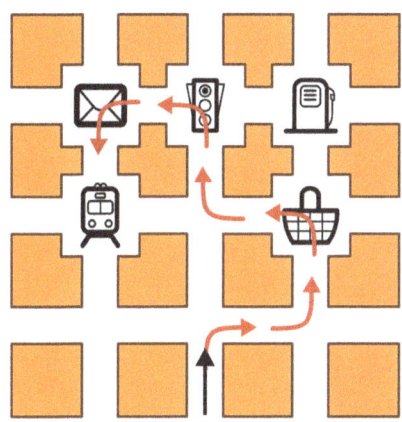

13 (D) Dannie
Betty and Andy were both born in the same month, which happens to be in May. Andy and Cathie were born on the same day, so it must have been on the 12th. Andy was born in May, so Cathie was born April 12th. This means Dannie was born on February 20th. The person whose birthday comes first is the oldest, so Dannie is the oldest.

14 (E) 5
Each of the 30 children at Adventure Park participated either in the "moving bridge" contest, went down the zip-line, or took part in both events. 20 children went down the zip-line, so 30 − 20 = 10 did not go down the zip-line but definitely participated in the "moving bridge" contest. There are 15 − 10 = 5 other children who participated in the "moving bridge" contest, so these 5 children took part in both events.

15 (B)

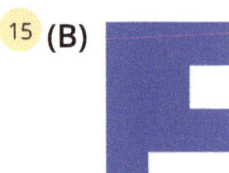

Piece (B) fits with the original piece as shown.

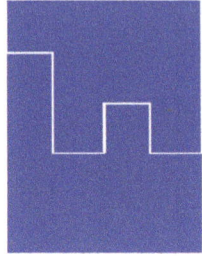

16 (B) 3

22 is divisible by 2, 24 is divisible by 4, and 25 divisible by 5. There are 3 numbers greater than 21 and smaller than 30 that have this property.

5 Point Solutions

17 (D) 16

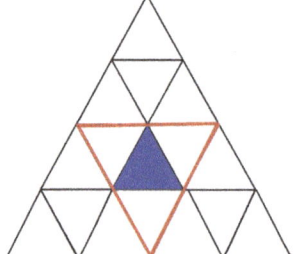

After the first step, we get the smaller red triangle and 3 others of the same size together covering the original triangle. Repeating the same steps with the red triangle we get an even smaller blue triangle and 3 other triangles of that size; together they cover the red triangle. We can do this to all the other triangles of the size of the red one, so the number of all the triangles of the size of the smallest resulting triangle (the blue one) that fit in the original triangle is 4 × 4 = 16.

18 (D) 102

Each number from 2013 to 2110 has at least one 0 as a digit, so the product of the digits is always 0 and is smaller than the sum of digits (the sum is greater than 2). 2110 is followed by 2111, 2112, 2113, 2114, and 2115. 2115 is the first year that has the product of digits greater than the sum of digits since 2 × 1 × 1 × 5 = 10 is greater than 2 + 1 + 1 + 5 = 9.
2115 − 2013 = 102, so 102 years will pass before the product of the digits in the notation of the year is greater than the sum of these digits.

19 (B) (31 − 7 × 3) × 24 × 60

There are 31 days in December and she slept 3 weeks or 7 days × 3, so the number of days she was awake is (31 − 7 × 3) days. Each day has 24 hours and each hour has 60 minutes, so she was awake for (31 − 7 × 3) × 24 × 60 minutes.

20 (C) 5

Among Basil's seven tiles there are the following squares: 3 single dot, 3 two-dot, 3 three-dot, 3 four-dot, 1 five-dot, and 1 six-dot. The tiles can only be arranged side by side if the numbers of dots on neighboring squares can be paired up. There are 4 such pairs of connecting tiles, one pair for each single dot, two-dot, three-dot, and four-dot squares, so the largest number of tiles Basil can arrange is 5. One of the possible arrangements is shown below.

Leftover domino tiles: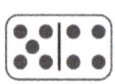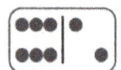

21 **(E) Such division is not possible.**
The price for all ten glass bells is 1 + 2 + 3 + 4 + 5 + 6 + 7 + 8 + 9 + 10 = 55 dollars. 55 is not a multiple of 3, so Cristi cannot divide all the glass bells into three packages so that each of the packages has the same price.

22 **(B) 67**
There are 9 squares along the width of 36 inches, so one square has the size of 4 × 4 inches. There are 15 squares along the length of 60 inches since 15 × 4 = 60. There are 15 vertical stripes (columns) in the fully unrolled rug. The first column has 4 moons, the second 5, the third 4, and it repeats itself so that 8 columns have 4 moons, and 7 columns have 5 moons. The total number of moons is 8 × 4 + 7 × 5 = 67.

23 **(B) 3**
Baby Roo can write one-digit, two-digit, three-digit, and four-digit numbers using only 0 and 1.
He must have used at least 3 of those numbers, by adding 1 three times, to get the ones digit of 2013. The only two possible sums of numbers Baby Roo could have written down are either
1 + 1001 + 1011 or 11 + 1001 + 1001.

24 **(B) 4**
The gray piece consists of 9 small squares but itself is not a square. Two gray pieces consist of 18 small squares and three gray pieces consist of 27 small squares. No square can be made of either 18 or 27 identical small squares, because the number of squares is also the lengths of the sides multiplied by each other, and these number can't be made by multiplying a number by itself. Four gray pieces consist of 36 small squares and the square of the size 6 × 6 also consists of 36 small squares, so there is a chance that using four gray pieces we can build a square. It can be done as shown here, so at least 4 gray pieces are needed to make a completely full gray square.

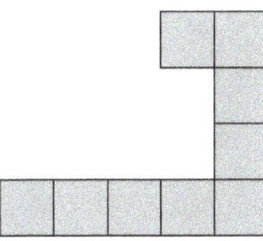

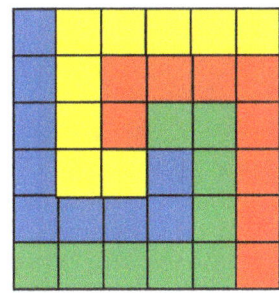

2015

2015

3 Point Solutions

1 (E) 15

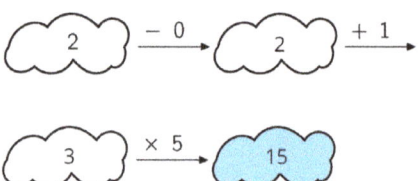

---

2 (A) A

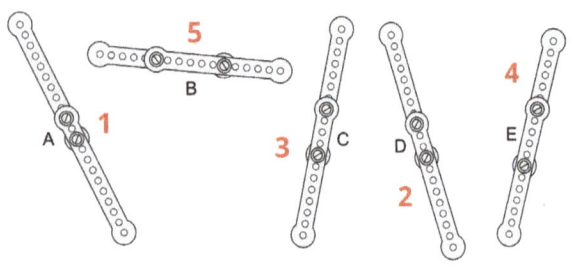

The longest connected strip Eric made was the one with the smallest number of holes between the two screws. The number of holes between the screws is shown next to each long strip. Strip A is the longest.

---

3 (E) 6

The number hidden behind the triangle is 3 since 3 + 4 = 7.
The number hidden behind the square is 6 since 6 + 3 = 9.

---

4 (C) (1000 − 1) ÷ 9
Evaluate each expression.
(A) (1000 − 100) ÷ 10 = 900 ÷ 10 = 90
(B) (1000 − 10) ÷ 9 = 990 ÷ 9 = 110
(C) (1000 − 1) ÷ 9 = 999 ÷ 9 = 111
(D) (1000− 100) ÷ 9 = 900 ÷ 9 = 100
(E) (1000 − 10) ÷ 10 = 900 ÷ 10 = 90
111 is the largest number, so (1000 − 1) ÷ 9 has the greatest value.

---

5 (E)

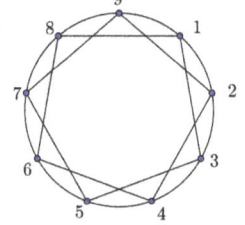

The dots are connected in the following order 1→ 3 → 5 → 7 → 9 → 2 → 4 → 6 → 8. All dots are connected in the pattern which is shown in figure (E).

---

6 (E) 8
Since the product of the two digits is 15, the only possibility for these two digits is 3 and 5. The sum of the digits is 3 + 5 = 8. It does not matter whether the number is 35 or 53.

---

7 (B) 6
The palm tree is growing on the island. Shade the land area connected to the palm tree and count the frogs in the shaded area. There are 6 frogs sitting on the island.

---

8 (D) Saturday
It will be Thursday again in 7, 14, 21, and 28 days. In 29 days it will be Friday, and 30 days from this Thursday it will be Saturday.

4 Point Solutions

9 (A)

KANGAROO is spelled clockwise (looking from above) on the umbrella, so the only possible three-letter pieces of the umbrella are: KAN, ANG, NGA, GAR, ARO, ROO, OOK and OKA. The only match is NGA in the figure (A).

---

10 (D) 15

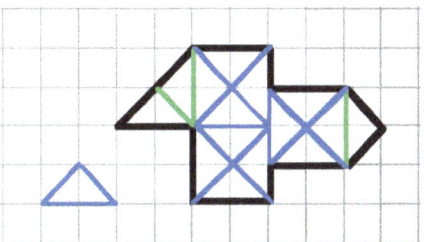

15 identical non-overlapping triangles of the given size are covering the shape.

---

11 (B) 3

After giving 2 apples to Yuri, Luis had 7 − 2 = 5 apples left. Luis also had 2 bananas, so Yuri gave Luis 3 bananas, since 2 + 3 = 5, and now Luis has 5 of each fruit.

---

12 (B) 4

Each grandchild received 4 pieces of candy and Grandma had 2 pieces left. With 2 more pieces she would be able to give each grandchild 1 more piece of candy. 2 + 2 = 4, so with 4 extra pieces of candy, giving 1 more piece to each grandchild means that Grandma has 4 grandchildren. Check: 4 × 4 + 2 = 18 and 4 × 5 − 2 = 18.

---

13 (C) 4

Not counting Tom, there are 9 skaters. The number of skaters who came after Tom is 3 more than the number of skaters who came before him. Removing those 3 extra skaters leaves 9 − 3 = 6 skaters with equal numbers before and after Tom. Thus, 3 skaters finished the competition before Tom. Tom ended up in 4th place.

---

14 (B) 4

Since both the ship and the airplane have to be next to the car, there are two possible arrangements for these 3 toys: either Ship, Car, Airplane or Airplane, Car, Ship. The ball can be placed either to the left or to the right of these two arrangements, giving us 4 possible ways in which the toys can be placed:

Ball, Ship, Car, Airplane,
Ball, Airplane, Car, Ship,
Ship, Car, Airplane, Ball,
Airplane, Car, Ship, Ball.

---

15 (D) D

Draw over the picture following the instructions for making turns. Notice that Peter never passes point D.

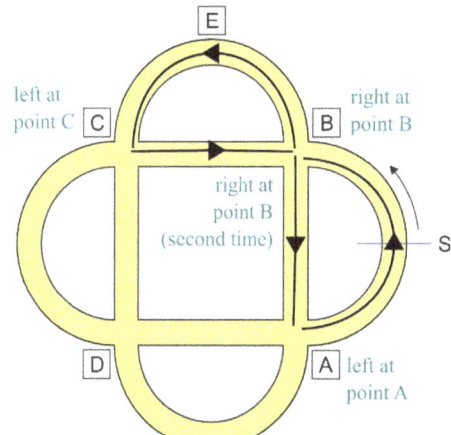

16 **(C) 6**

There are 5 ladybugs: one ladybug with 2 spots, two ladybugs with 3 spots, one ladybug with 5 spots, and one ladybug with 6 spots. Since each ladybug sent one text to a ladybug whose number of spots differs exactly by 1, it means that the texts were exchanged between 2 spotted and 3-spotted ladybugs, and separately between 5-spotted and 6-spotted ladybugs. The ladybug with 2 spots sent one text to each of the two ladybugs with 3 spots (2 texts) and received a text back from both (2 texts). The ladybug with 5 spots sent a text to the ladybug with 6 spots and received a text back (2 texts). Altogether 6 text greetings were sent.

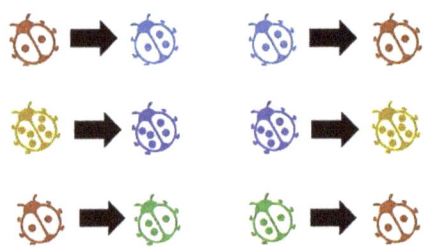

5 Point Solutions

17 **(A)**

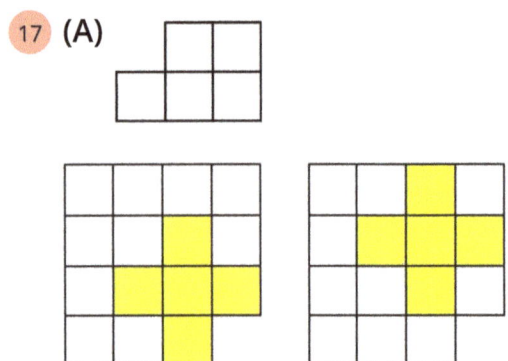

Piece (E) fits nicely in the cut off corner but no other corner of the figure can be covered by (E).

The upper right corner can be covered by piece (D) only when the longer side of (D) is horizontal. The lower left corner can be covered by piece (D) only when the longer side of (D) is vertical. One piece (D) in a horizontal position and another one in a vertical position always overlap, so the figure cannot be covered by non-overlapping pieces (D).

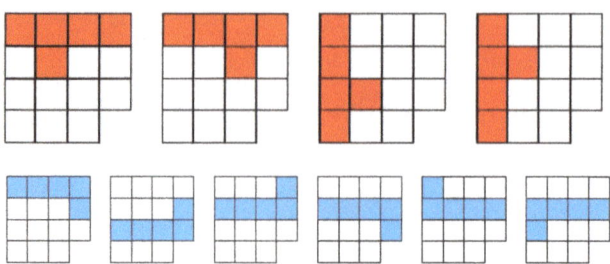

The middle square of the shorter column can be covered by piece (C) when the longer side of it is placed horizontally. By symmetry, the middle square of the shorter row can be covered by piece (C) when the longer side of it is placed vertically. One piece (C) in a horizontal position and another one in a vertical position always overlap, so the figure cannot be covered by non-overlapping pieces (C).

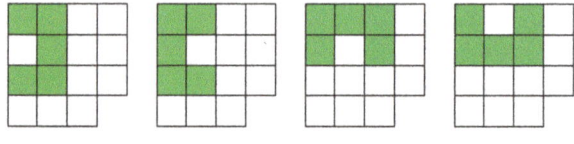

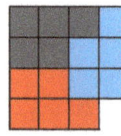

The upper left corner of the figure can be covered in four different ways by piece (B). The same is true for the lower left and the upper right corners.

In the first two cases shown above any coverage of the lower left corner overlaps each coverage of the upper left corner. In the next two cases any coverage of the upper right corner overlaps each coverage of the upper left corner, so the figure cannot be covered by non-overlapping pieces (B). As shown to the right, the figure can be divided into three identical pieces that look like (A). We can use symmetry to get another arrangement.

18. (C) 13

A cube has 4 edges in each of the three directions, so it has 12 edges altogether. Along each edge there is exactly one small white cube, so there are 12 small white cubes along all edges. Another small cube is at the center of the big cube. It is white since it is adjacent to the gray middle cubes of the six faces. Thus, Jack used 12 + 1 = 13 white cubes.

19. (A) 4

Let's call the small pitcher S, the medium pitcher M, and the large pitcher L. There are two ways to fill in the barrel with water: the first option is to use 6S + 3M + 1L, the second option is to use 2S + 1M + 3L. A clever way to approach this problem is to notice the proportion of small and medium pitchers for the two options of filling the barrel. In the first option it is 6S + 3M and in the second option it is 2S + 1M. If we triple the number of small and medium pitchers in the second option, 3 × (2S + 1M), we get a quantity equivalent to the first option (6S + 3M). Let's consider what it would take to fill 3 barrels. Using the second option 3 × (2S + 1M + 3L), it would take 6S + 3M + 9L to fill 3 barrels, which we can express as (6S + 3M + 1L) + 8L. From the first option, 6S + 3M + 1L is equivalent to 1 barrel, so 8 large pitchers are equivalent to 2 barrels. To fill a barrel, we would need 4 large pitchers.

20. (D) 5 or 7

After picking one out of the 5 given numbers (2, 3, 5, 6, and 7) to be placed in the center square, the remaining 4 numbers must form pairs where the sum of the smallest and the largest is equal to the sum of the other two. We can eliminate both even numbers (2 or 6) from being placed in the center square, since the remaining 4 numbers would consist of one even and three odd numbers, which could not form equal sums. Placing 3 in the center square would not work, since 2 + 7 ≠ 5 + 6. Placing 5 in the center square makes the sums 2 + 7 and 3 + 6 equal. Placing 7 in the center square would also work, since 2 + 6 = 3 + 5. In conclusion, either 5 or 7 can be placed in the center square of the cross.

21. (E) 15

For the product to equal to 0, one of its factors must be a 0. This is one of Peter's three numbers.

The product of Ann's, George's, and Peter's numbers (without 0) is the product of all digits from 1 to 9. That product is 1 × 2 × 3 × 4 × 5 × 6 × 7 × 8 × 9 and, with no references to balls, can be written as 2 × 4 × 7 × 3 × 5 × 6 × 8 × 9 = 56 × 90 × 72. 90 is the product of Ann's numbers and 72 is the product of George's numbers, so 56 must be the product of Peter's two nonzero numbers less than 10. The only option for Peter is 56 = 7 × 8, so the sum of his three numbers is 0 + 7 + 8 = 15. For Ann and George there are two options 90 = 9 × 5 × 2 and 72 = 6 × 4 × 3 × 1 or 90 = 6 × 5 × 3 and 72 = 9 × 4 × 2 × 1.

22 (C)

Assign numbers from 1 to 6 (from left to right) to the ends of the ropes. For the original three ropes the ends are 1 & 4, 2 & 6, and 3 & 5.

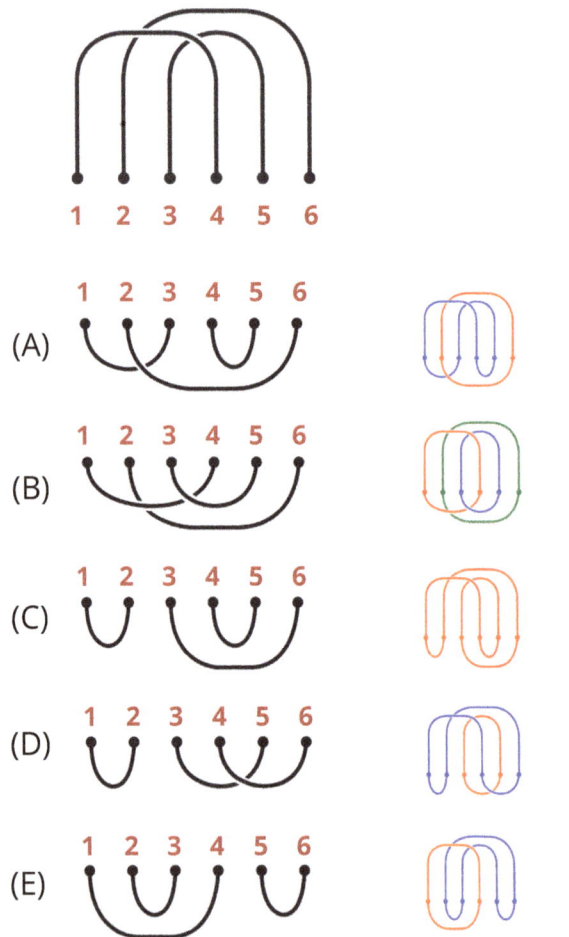

Set (A) will create a short closed loop with the ends 2 & 6. Set (B) will create three short closed loops with the ends 1 & 4, 2 & 6, and 3 & 5. Set (D) will create a short closed loop with the ends 3 & 5. Set (E) will create a short closed loop with the ends 1 & 4. Only set (C) creates a loop that will make one complete rope.

23 (D) 8

Keep the top square as it is and rotate the other two squares. There are 4 different configurations of the middle square obtained by its rotations shown in a row below.

Due to the symmetry, there are only 2 different configurations of the bottom square obtained by its rotations shown in a column below.

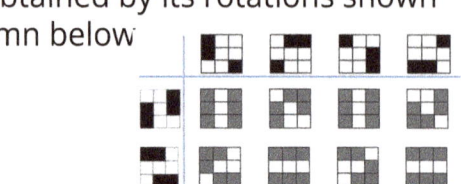

There are 2 × 4 = 8 squares when the two column configurations are put on the top of the four row configurations. Below the same matches are displayed when the top square is added.

The maximum number of black squares is 8.

24 (C) Charlie

Anna made 6 times as many cookies in two days as she made on Friday, since among the numbers 24, 25, 26, 27, and 28, only 24 is a multiple of 6, so Anna made 4 cookies on Friday. Berta made 5 times as many cookies in two days as she made on Friday, since only 25 is a multiple of 5, so Berta made 5 cookies on Friday. Elisa made 4 times as many cookies in two days as she made on Friday, since only 24 and 28 are multiples of 4, so Elisa made 7 cookies on Friday. David made 3 times as many cookies in two days as he made on Friday, since only 24 and 27 are multiples of 3, so David made 9 cookies on Friday.
Finally, Charlie made twice as many cookies in two days as he made on Friday, so Charlie made 13 cookies on Friday. Charlie baked the most cookies on Friday.

2017

3 Point Solutions

1 (E)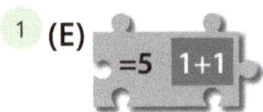

Notice that 8 – 3 is 5. That means that the puzzle we are looking for has to start with = 5. This eliminates answers (B) and (D). Also the second addition or subtraction must be equal to 2. This eliminates (A) and (C). The only puzzle piece fitting between the two given puzzles is puzzle (E).

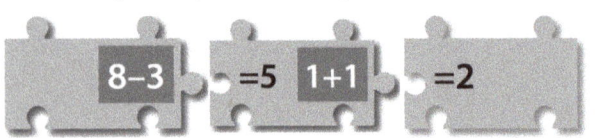

2 (A) 12

In the picture there are 6 kangaroos. John sees only half of the kangaroos. The other half must also be 6. The total number of kangaroos is 12 since 6 + 6 = 12.

3 (E)

Any of the pictures cannot be seen if they are covered with at least one black square. Sliding both transparent grids with some dark squares would result in all but one square not covered with at least one black square. That square is in the top row, middle column, showing the butterfly.

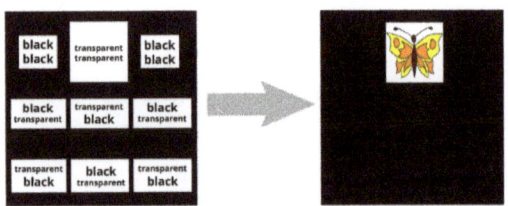

4 (C)

All the sets of footprints but one from the first picture repeat in the upside-down picture. The picture highlighted in yellow does not appear in the upside-down image.

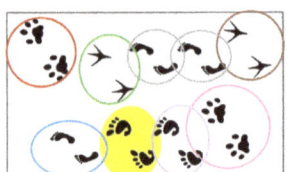 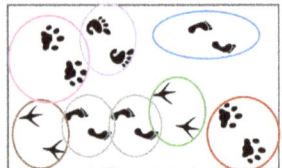

5 (A) 16

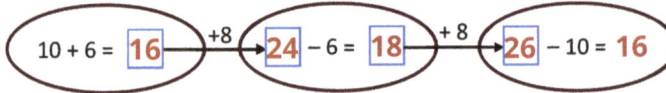

Complete the addition and subtraction statements from left to right. 26 – 10 = 16 is the number hidden behind the panda.

6 (E) 16

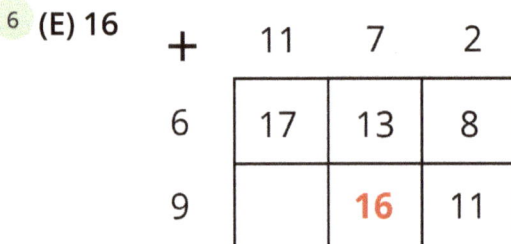

The sums in the upper row are obtained by adding 6 to the numbers above the grid. They are 17 = 11 + 6, 13 = 7 + 6, and 8 = 2 + 6. The numbers in the lower row are obtained by adding 9 to the numbers above the grid since 11 = 2 + 9. The number in the box with the question mark is 7 + 9 = 16.

7 (C) 4

There are 4 pieces that have exactly four sides. They are marked with red stars in the picture. All remaining broken mirror pieces have either three or five sides.

8 (A)

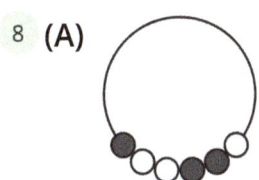

The position of the necklace in its original form is shown here:

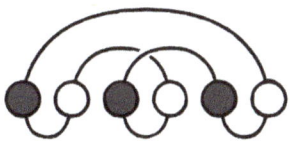

The position of the necklace after untangling the middle two beads is shown.

The position of the beads that are connected on the necklace is shown in (A):

4 Point Solutions

9 (E)

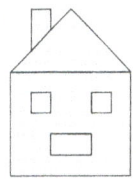

Notice that standing in the back of the house the chimney would be seen on the left side of the roof. Additionally, the house has three windows and no door. Houses (A) and (B) have doors. Houses (C) and (D) have chimneys on the right side of the house. Only house (E) has a chimney on the left side of the house, no door, and three windows.

10 (E) ● + ● = ■

The statement

can be simplified by removing one square from each side, so

● + ● + ● + ● = ■ + ■

As indicated by parentheses below, there are two identical pairs on the left side and two identical squares on the right side,

so one pair matches one square as shown here.

● + ● = ■

Therefore, only statement (E) is true.

11 (B) 4

Getting three 25-balloon packets is more than 70 balloons since 3 × 25 = 75. Any other combination of 3 packets, such as two 25-balloon packets and one 10-balloon packet, is not enough since 2 × 25 + 1 × 10 = 60 is less than 70 balloons. Marius can buy exactly 70 balloons by getting 4 packets, two 25-balloon packets and two 10-balloon packets, 2 × 25 + 2 × 10 = 70.

12 (C)

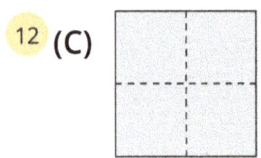

Folding the paper as shown in (C) produces the desired result:

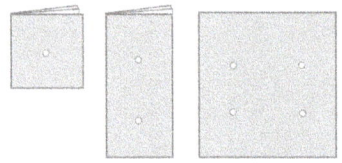

Folding the paper in other ways will result in different patterns as shown below:

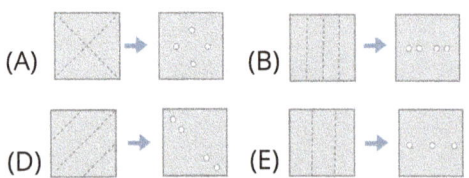

13 (D) 4

13 + 19 = 32 children signed up for the tournament. 32 children cannot be divided into 6 teams evenly. 36 is the first number greater than 32 that is divisible by 6. At least 4 more children (36 − 32 = 4) need to sign up to form six teams with an equal number of members each.

14 (D) 14

1	2	1	3
4	1	1	2
1	7	3	2
2	1	3	1

There are three 2 × 2 squares if we look at rows 1-2 together. The same is true looking at rows 2-3 and at rows 3-4. The sums for the 2 × 2 squares are: 8, 5, and 7 in rows 1-2; 13, 12, and 8 in rows 2-3; 11, 14, and 9 in row 3-4. The largest sum is 14 that is obtained by adding 7, 3, 3, and 1.

15 (C) 75 minutes

The time needed to cook all 5 dishes is 40 + 15 + 35 + 10 + 45 = 145 minutes. Each cooking time is a multiple of 5, so the optimal times must be multiples of 5. The shortest cooking time would occur if the total cooking time of all 5 dishes could be equally divided among two burners. This cannot happen because 145 ÷ 2 = 72 1/2 and it is not a multiple of 5, so we must look for the total cooking time to be close to half the time needed to cook all 5 dishes and to be a multiple of 5. The optimal distribution of the total cooking times is 75 + 70 and it can actually happen. On one burner David can cook dishes with the cooking times of 40 and 35 minutes (75 minutes total) and on the second burner dishes with the cooking times of 15, 10, and 45 minutes (70 minutes total). Both burners would work at the same time. The shortest time needed to cook all five dishes is 75 minutes.

16 (D) 13

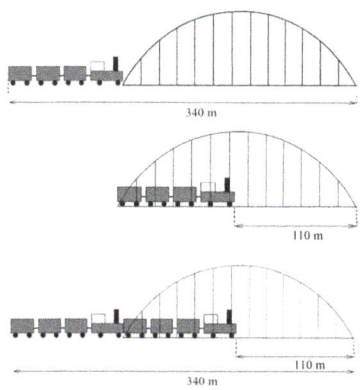

Look at the diagram and the operations to be performed. Any number multiplied by zero results in 0, so the number to the left of the question mark is 0. Follow the arrows to perform all other operations. The number 13 should be written in the circle with the question mark, which is marked in yellow here.

5 Point Solutions

17 (C) 5

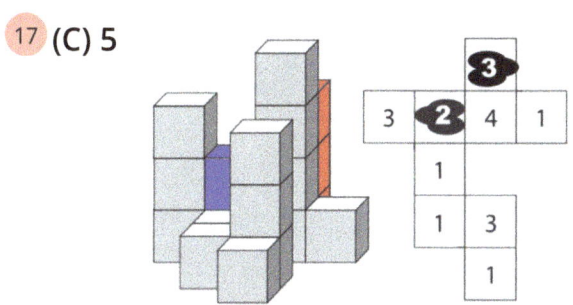

The plan represents the shape of the base of the building. The numbers represent how many blocks are put on top of each square space. There are 3 blocks shown by the upper square covered by ink. They are marked in red in the figure. There are 2 blocks shown by the other square covered by ink, one shown in blue and one underneath. The sum is 3 + 2 = 5.

18 (B) 115 m

Overlaying the two given pictures gives a useful presentation of this problem.

340 meters is the sum of the lengths of the bridge and the train.
When the train is on the bridge, the remaining length of the bridge is 110 meters. As shown in the last picture, 340 meters is the sum of the lengths of two trains and 110 meters. The length of two identical trains is 340 meters – 110 meters = 230 meters. The length of the train is 230 meters ÷ 2 = 115 meters.

19 (E) X

John was born in February and the problem asks about how old he is in March, so his birthday has already passed. He was born in the year MMVII. From the question's explanation, M=1000, the other M=1000, V=5, I=1, and I=1. Adding the numbers together we get: 1000 + 1000 + 5 + 1 + 1 = 2007. In order to calculate how old John was in 2017, it is enough to subtract the two given years. 2017 – 2007 = 10. John was 10 years old, which is X in Roman numerals.

20 (D) 9

Susan can start her tour with the giraffe, the elephant, or the turtle since she does not want to start with the lion. She wants to see two different animals. After starting with the giraffe, she can then see the elephant, the turtle, or the lion. After starting with the elephant, she can then see the giraffe, the turtle, or the lion. After starting with the turtle, she can then see the giraffe, the elephant, or the lion. Thus, she can start with 3 different animals and then for each one she has 3 choices at to the second animal she sees. So, there are 9 different tours she can plan. The table below shows Susan's nine options to see two different animals.

21 (C) 5

Start with the fact that three brothers ate 9 cookies, so the fourth brother ate 11− 9 = 2 cookies. Among the three brothers, one ate 3 cookies, so the two remaining brothers together ate 9 − 3 = 6 cookies. As a sum of two different numbers (ignoring the order) 6 is either 1 + 5 or 2 + 4. The fourth brother ate 2 cookies, so the only valid option left is 1 + 5 = 6. 5 was the largest number of cookies that one of the brothers ate.

22 (B) 5

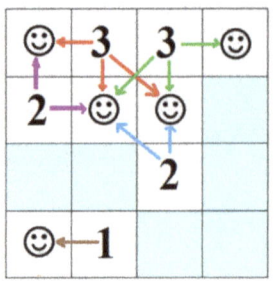

In order to determine where the smileys are hiding, start with selecting the order of discovering cells with a smiley. Step 1 – RED arrows indicate the only options of 3 neighboring cells with a smiley; Step 2 – BLUE and PURPLE arrows are determined by Step 1, so there are no more smileys around two cells with 2; Step 3 – two GREEN arrows are already determined by Step 1 and there is only one option for the third smiley; Step 4 – for the BROWN arrow there is only one option left in the lower left corner. The cells left shaded cannot have smileys. Zosia hid 5 smileys as the diagram shows.

23 (E) 19

The total number of pieces of candy in the bags is 1 + 2 + 3 + 4 + 5 + 6 + 7 + 8 + 9 + 10 = 55. The first four boys took 5 + 7 + 9 + 15 = 36 pieces of candy, so Eric got 55 − 36 = 19 pieces of candy.

24 (B) 2

The total number of petals is 6 + 7 + 8 + 11 = 32 = 30 + 2. After 10 rounds of tearing off one petal from each of any three flowers (if it can be done), Kate will stop with 2 petals left. It can be done in many different ways. One is shown in the picture.

2019

2019

3 Point Solutions

1 (E)

The runner with D on the shirt stands on the highest step. The next one is B and the third one is E.

2 (C)

12 is 10 + 2, which is the same as 5 + 5 + 1 + 1, so the picture will have two bars and two dots. This is picture (C).

3 (A) Tuesday

Since yesterday was Sunday, today is Monday and tomorrow will be Tuesday.

4 (D)

Olaf sees the second, fourth, and fifth vehicle (counting from the left) when he closes his book. The distances from the book spiral to the first opening is 3 squares, which will cover up the red car. The two by two opening would show the blue motorcycle. The distances from the book spiral to the second opening is 7 squares, so the next vehicle, which is the green truck, is covered. The four by two opening would show the orange SUV and the purple tractor. This is shown in answer (D).

5 (A)

The only piece that Karina can get is the one with a star and a club shown in the middle of the top row. All other pieces are either not next to each other or the shapes don't face the right direction.

6 (A)

The track of shoeprints in the straight line from right to left is covered by the other two. This means this was the first person walking on the snow. The second track, which covers the first track but is covered by the third, starts in the lower right corner goes up and then down. The third is the track that is on top of the other two, moving from the upper right corner to the middle left. This order is shown in answer (A).

7 (D)

There are 10 connected sticks in Pia's shape shown in the picture. Answers (A), (B), (C), and (E) also show 10 connected sticks. Answer (D) shows a shape with 12 connected sticks, which is more than Pia has.

8 (B) 5

Starting with 2 + 1 = 3, 0 + 3 = 3, and 1 + 8 = 9 (answers shown in blue), we can calculate 8 − 3 = 5 (shown in red), which is the replacement for the question mark.

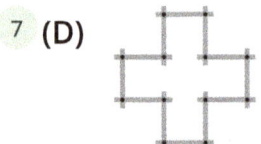

4 Point Solutions

9 (B) 16

Linda uses 8 pins to pin 3 photos on a cork board as shown in the picture. She uses four pins for the first photo, but because of overlapping edges, she only needs two more pins for each additional photo. When pinning 7 photos, Peter would use 16 pins (4 + 6 × 2 = 4 + 12 = 16).

10 (C) 3

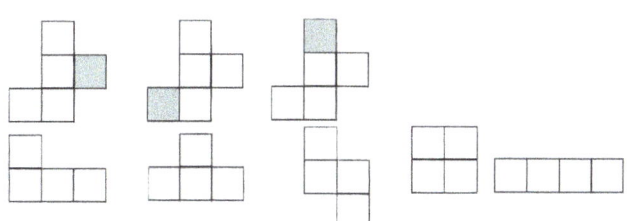

By removing one square, marked in gray, only the first, second, and third shape can be obtained. So, Dennis can get 3 out of 5 listed shapes.

11 (C)

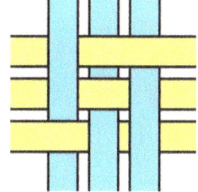

From the back, the middle vertical strip will be behind the upper two horizontal strips and the third horizontal strip will be behind the middle vertical strip. Only picture (C) shows this.
Also, the left vertical strip of (C) matches the right strip of the original pattern when looking from behind. The same is true for the right vertical strip of (C) and the left vertical strip of the original pattern.

12 (E) 11 kg

From the first scale we know the toy dog's weight is less than 12 kg. From the second scale we see that 2 dogs weigh more than 20 kg, which means one dog weighs more than 10 kg. There is only one whole number that is greater than 10 but less than 12, and it is 11.

13 (B) 10

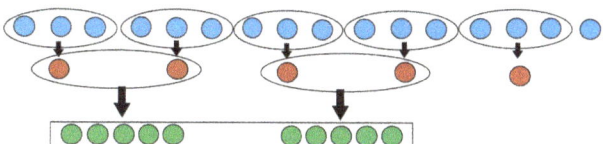

Sara will first need to trade blue marbles for red marbles, and then red marbles for green ones.

14 (A) Either 0 or 1

In order to get the largest possible sum, Steven needs to put the greatest digits first, so the hundreds digit is 9 and the tens digit is 2. Steven will add the last digit of the 3-digit number to the other remaining digit, so it does not matter whether he places 1 or 0 there. The sum will be the same.

15 (D) 250

Since the glass full of water weighs 400 grams and the empty glass weighs 100 grams, the water alone weighs 400 − 100 = 300 grams. A glass half-full would weigh 100 grams (glass only) + 150 grams (half of 300 grams of water), which gives the sum of 250 grams.

16. (D) 11 cents

22 cents for 2 apples, 2 pears, and 2 bananas

By combining all three costs and rearranging the fruit we can conclude the following:

If 2 apples, 2 pears, and 2 bananas cost 22 cents, then 1 apple, 1 pear, and 1 banana cost half of this, which is 11 cents.

5 Point Solutions

17. (E) 6

Starting with the middle row we can calculate that the circle stands for 4, as 12 divided by 3 is 4. This gives us the following:

We can subtract 4 from the first row to get the following:

Notice that one more heart in the last row makes the sum larger by 5, so the heart stands for 5. Now we know that the star stands for 6, because 6 + 5 = 11 and 6 + 5 + 5 = 16.

18. (C) 44

Anna used a set of 7 squares along each of the four sides of the picture and added four more squares at each corner. (They are colored in the picture.) A picture that is 10 by 10 requires the frame to have 10 squares along each side plus one in each of the four corners.
10 + 10 + 10 + 10 + 4 = 44 squares.

19. (B) 64

From 1 to 49 the digit 5 appears 5 times and from 50 to 59 it appears 10 + 1 = 11 more times. Thus, from 1 to 59 the digit 5 appears 5 + 11 = 16 times as given. We can have more pages included before the digit 5 occurs for the 17th time on page 65. Thus, the maximum number of pages this book can have is 64. The digit 5 that appears 16 times is marked in red and the last number represents the maximum number of pages.

1, 2, 3, 4, 5, 6, 7, 8, 9, 10, 11, 12, 13, 14, 15, 16, 17, 18, 19, 20, 21, 22, 23, 24, 25, 26, 27, 28, 29, 30, 31, 32, 33, 34, 35, 36, 37, 38, 39, 40, 41, 42, 43, 44, 45, 46, 47, 48, 49, 50, 51, 52, 53, 54, 55, 56, 57, 58, 59, 60, 61, 62, 63, 64.

20 (E) 83

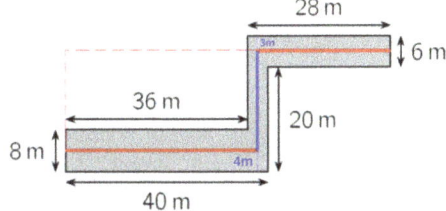

The horizontal length of the path is equal to the sum of 36 and 28, which is 64. The red line in the picture marks the horizontal path. The vertical length of the path is equal to the sum of (20 − 4) and (6 − 3), which is 16 + 3 = 19. The blue line in the picture marks the vertical path. The cat walks 64 + 19 = 83 meters.

21 (B) 3

Since 10 of the animals are not cows, we can conclude that there are 5 cows, since there are 15 animals total. Since 8 are not cats, we can conclude that 15 − 8 = 7 are cats. This leaves us with the number of kangaroos in the park as 15 animals − 5 cows − 7 cats = 3 kangaroos.

22 (E) 1 and 3 are yellow

Triangle 4 must be blue as it has an edge common with red and yellow. Triangle 5 must be red as it has an edge common with yellow and blue.

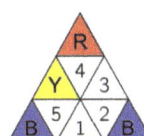 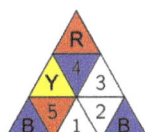

For triangles 1, 2, and 3, Mary must use the remaining triangles, two of which are yellow and one is red. Since two triangles of the same color cannot share an edge, the yellow ones must be 1 and 3. This leaves us with only one option, red, for triangle 2.

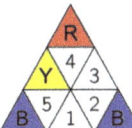 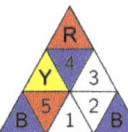 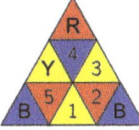

Answer "(E) 1 and 3 are yellow" is the only one that correctly describes the figure.

23 (B) Bartek

If Edek is telling the truth, then Alek ate the cookie, and Alek would be a liar by contradicting Edek. There is only one liar among the five boys, so Bartek's statement would be true. Bartek says that he (not Alek) ate the cookie. Therefore, Edek is a liar and everybody else is telling the truth, so Bartek ate the cookie.

24 (D) 22

At the beginning hang all 35 towels as in figure 2. The number of pegs still available is 58 − (1 + 35) = 22 since each towel has one peg at its right end (35 pegs) and the leftmost towel also has one peg at its left end. One by one, we move 22 consecutive towels to the left. Each of these 22 towels already has a peg at its left end and we add a new peg to the right end of each towel. Therefore, Emil hung up 22 towels as shown in figure 1.

Notice that 35 − 22 = 13 towels were hung up as shown in figure 2.

Altogether, 22 × 2 + (1 + 13) = 44 + 14 = 58 pegs were used by Emil.

figure 1

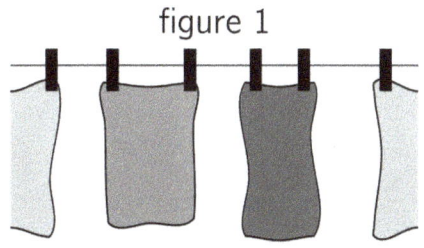

figure 2

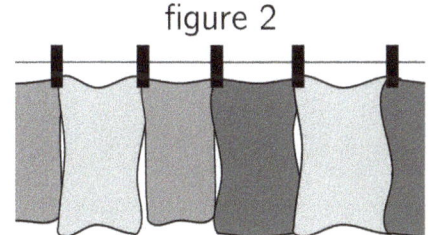

2021

2021

3 Point Solutions

1 (C)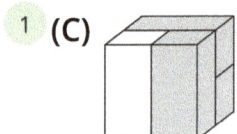

There is only one white brick. The only cube with one white brick is cube (C). The rest have two white bricks, so they are impossible to make.

---

2 (C) 6
We start at the ring and follow the line. The first, third, sixth, seventh, eighth, and tenth fish are pointing towards the ring, for a total of 6 fish, which is the answer (C).

---

3 (B) 15

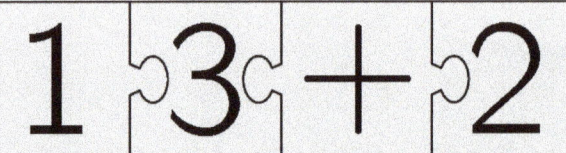

The calculation formed by the pieces is "13 + 2." The result is 15, which is the answer (B).

---

4 (B)

Piece (B) is on the left side of the picture of the sun. None of the other pieces are there.

---

5 (E)

Add the numbers in the rings with the dots (marks from the shots) to get the scores. It's possible to score the same value more than once.

For (A), the score is 10 + 8 + 7 = 25. For (B), it is 10 + 7 + 7 = 24. For (C), it is 8 + 8 + 8 = 24. For (D), it is 8 + 7 + 7 = 22. For (E), it is 9 + 9 + 8 = 26.

The most points scored was 26 on target (E), so this was Ricky's target.

---

6 (C) 48
The numbers 6 and 27 are directly below the question mark. To get from 6 to 27, we added 21, so 21 is the distance around the cylinder. To get to the question mark from 27, the tape goes around the cylinder again, so we need to add 21. So. the number at the place shown with the question mark is 21 + 27 = 48.

---

7 (D) 13
We need to find two numbers that differ by 6 and add up to 20. These numbers are 7 and 13. The larger number, 13, is how many stars the gold rocket exploded into.

8 **(C) 5 kg**
The first scale tells us that one black ball and one gray ball weigh 6 kg. We can remove one black ball and one gray ball from the third scale to see that one gray ball weighs 4 kg. We can remove the gray ball from the middle scale to see that two white balls weigh 10 kg, so one white ball weighs 5 kg.

4 Point Solutions

9 **(A)**

It is impossible to get the same fruits together in set (A). There are two apples and two grapes. To get all the apples together and all the grapes together, we need to move an apple and a grape. We can either swap cards 1 and 3, 1 and 5, 3 and 4, or 4 and 5, but none of the options work.

For set (B), we can swap cards 1 and 4. For set (C), we can swap cards 1 and 5. For set (D), we can swap cards 3 and 5. For set (E), we can swap cards 3 and 4. Therefore, set (A) is the only one where it is impossible to get the same fruits together.

10 **(E)**

Sofie must pick the circle from container 3. This forces her to take the star from container 1. This again forces her to take the pentagon from container 5. Since the circle, star, and pentagon are already taken, she must take the diamond from container 4.

11 **(E)**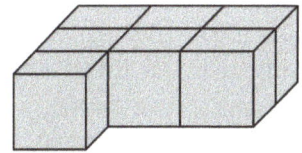

There are 18 cubes in total. The white piece has 4 cubes and the black piece has 7 cubes.

There are 7 cubes remaining. The only gray piece with seven cubes is piece (E).

12 **(C)**

First, the 9 and 3 balls combine into a 12 ball going to the right. Then, the 12 and 7 balls combine into a 19 ball going to the right. Then, the 19 and 20 balls combine into a 39 ball going to the left. Finally, the 39 and 10 balls combine into a 49 ball going to the left.

13 **(C) 40**
The shop gained 120 – 70 = 50 dollars when it moved from having sold 6 ice cream cones to having sold 16 ice cream cones. So, it sold 10 cones to make 50 dollars, and now we can see that one cone cost 50 ÷ 10 = 5 dollars. The shop gained 6 × 5 = 30 dollars when it sold the first 6 cones, so before that there were 70 – 30 = 40 dollars in the drawer.

14 **(E) 38**
If we put the number of leaves left on the first branch with the number of leaves left on the second branch we get 20, because whatever number the Koala left on the second branch is 20 minus that number for the leaves left on the first branch. 20 minus a number plus that same number is 20. Then there were 20 − 2 = 18 leaves left on the third branch. Together, there were 20 + 18 = 38 leaves left.

15 (D) 20

The total height is 48, given by the longest ladder. The overlap between the 32 and 36 ladders is the same height as the shortest ladder. If we add 32 + 36 = 68, the shortest ladder gets counted twice. Since the total height is 48, the height of the shortest ladder is 68 − 48 = 20.

16 (B)

The first three moves flip all of the cups over. The next three flip them back, and the next three flip them over again. At this point, the cups look like:

In the tenth move, we move the left cup to the right and flip it, to get picture (B).

5 Point Solutions

17 (D) 4

Eva puts the apple on square 1. The flower, circle, and triangle are all next to each other, so they can either go on squares 2, 3, and 4, or squares 3, 4, and 5. If we put them on squares 2, 3, and 4, we have to put the star on square 5, which is not allowed. We must then put them on squares 3, 4, and 5. Since the flower is next to both the circle and the triangle, it must go in between them, so it is on square 4.

18 (E) G

The sum of the top is 7 + 5 + 4 + 2 + 8 + 3 + 2 = 31 and the sum of the bottom is 4 + 3 + 5 + 5 + 7 + 7 + 4 = 35. Since we want the same sum on the top and bottom, we want each row to add up to 33. We do this by removing 2 from the bottom row and adding 2 to the top row. The only card where the two numbers have a difference of 2 is card G, so our answer is (E).

19 (D) 7

The sum of all the numbers is 45. The sum of the last 8 numbers is 7 + 9 + 11 + 9 = 36. This means that the first number is 45 − 36 = 9. The second number is 15 − 9 = 6, the third number is 7 − 6 = 1. The fourth number is 3 − 1 = 2. Finally, the number in the shaded box is 9 − 2 = 7.

20 (A) 3
The only ways to get 30 points are 3 + 9 + 18 and 3 + 13 + 14. She must have hit the balloon worth 3 points.

21 (D) 36
Since the number of cookies can be evenly divided by 2, 3, and 4, it must be a multiple of 12. This means that the box has 12, 24, 36, or 48 cookies. Adding 6 more cookies gives 18, 30, 42, or 54 cookies. Out of these, only 42 cookies can be divided among 7 children, so there must be 36 cookies in the box.

22 (E) 1 and 4
The total weight is 7 + 5 + 6 + 2 + 16 = 36 kg. The bananas must weigh 27 kg and the apples must weigh 9 kg. The only boxes that add up to 9 kg are boxes 1 and 4.

23 (D) 6
8 is above the 7 and 9 is to the left of 8, because they both have to be bigger than 7. The center, bottom-middle, middle-right, and bottom-right boxes are all less than 5, so they must be 1, 2, 3, and 4 is some order. The only remaining number is 6, which has to go in the box with the question mark.

24 (A) 1 square
We can remove the hexagons from the middle scale to see that 1 triangle weighs the same as 5 squares. We can replace the triangle on the first scale with 5 squares to get that 2 hexagons weigh the same as 6 squares, or that 1 hexagon weighs the same as 3 squares. On the third scale, the left side weighs the same as 9 squares while the right side weighs the same as 10 squares, so to balance them, we need to put 1 square on the left side.

2023

3 Point Solutions

1 **(D) D**
The candles are identical, so they burn at the same rate. The more time a candle burns, the shorter it gets. The candle that went out first burned the least amount of time, so it is the tallest candle. This is candle D.

2 **(C) 5**
The coins marked with numbers add up to 20 + 10 + 10 + 1 = 41. The total value is 51, so the two coins with question marks together have a value of 51 − 41 = 10. They are both the same, so each one has a value of 5.

3 **(B) 5 and 9**
The picture shows that one hole will have a time 4 hours after the other. This is only true for the hours in (B), 5 and 9.

4 **(C) 1 and 4**

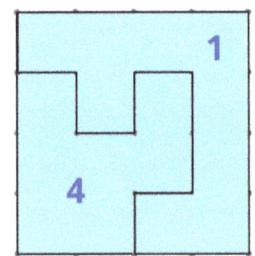

Use the dots on the grid to find the area of the square. It is 16. So, the areas of the two pieces we put together must add up to 16. Piece 1 has an area of 9, piece 2 has an area of 8, piece 3 has an area of 10, and piece 4 has an area of 7. The only two areas that give a sum of 16 are 9 + 7, which are pieces 1 and 4.

5 **(C) 8 minutes**

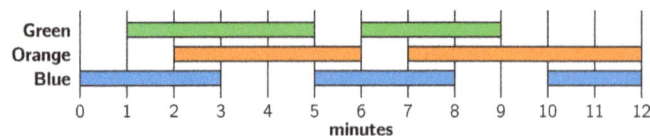

The diagram shows twelve places (minutes), numbered from 1 to 12, at which the lights are turned on or turned off. There are exactly 2 lights on during the following intervals (minutes): 1-2 (green and blue lights), which is 1 minute; 3-5 (green and orange lights), which is 2 minutes; 5-6 (orange and blue lights), which is 1 minute; 6-7 (green and blue lights), which is 1 minute; 8-9 (green and orange lights), which is 1 minute; and 10-12 (orange and blue lights), which is 2 minutes. Thus, there are exactly 2 lights on for 8 minutes.

6 **(E)**

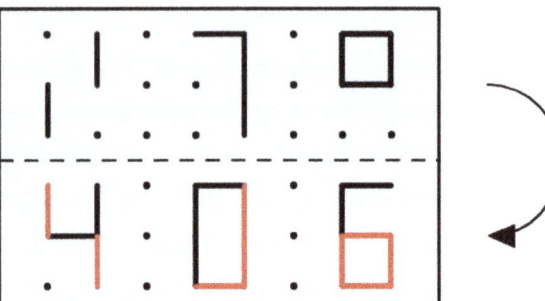

7 (C) 4

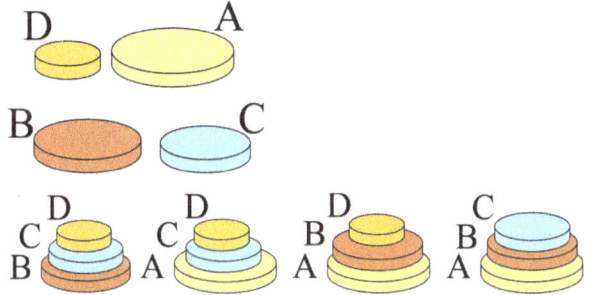

Let's label the largest disc A, the next largest B, the next C, and the smallest D. If Anna does not use disc A, she can make exactly one tower: BCD. If she does not use B, she can make exactly one tower: ACD. If she does not use C, she can make exactly one tower: ABD. And, if she does not use D, she can make exactly one tower: ABC. Therefore, Anna can make 4 different towers.

---

8 (E)

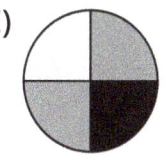 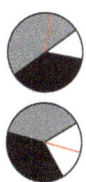

Danny can make figure (A) by gluing the gray piece over the black piece, and the white piece over the gray piece. He also can make figure (B) by gluing the white piece over the black piece, and the gray piece over the black and the white pieces, as shown.

It is possible for him to make (C) by gluing the white and gray pieces over the black piece.

Finally, it is possible to make (D) by gluing the white piece over the black piece, and the gray piece over the black and the white pieces, as shown.

However, it is impossible to make figure (E), because the gray piece is shaped like two white pieces together, while this option shows two gray pieces separated.

4 Point Solutions

9 (A)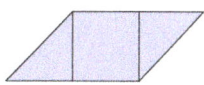

Piece (C) can only be placed in the corner.

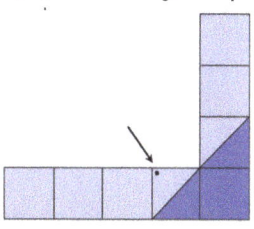

Now there are three and a half squares on the horizontal part and two and a half on the vertical part that still need to be covered.

Putting pieces (B) and (D) can make the figure

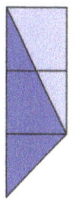

to cover the vertical part. Putting pieces (A) and (C) can make the figure

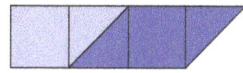

to cover the horizontal part. So, the piece that covers the dot is (A).

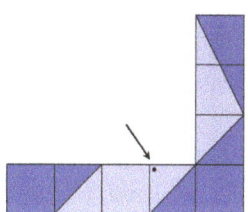

10 (A) 1 kg

The 5 kg and 6 kg weights are already used. If we set aside either the 2 kg or 4 kg weight, the total weight that we use will be an odd number (19 kg or 17 kg). With an odd number, it would be impossible to divide the weights so that the scales balance. So, Amy put aside either the 1 kg or 3 kg weight. Let's check which one works. Without the 1 kg weight, the total weight will be 2 kg + 3 kg + 4 kg + 5 kg + 6 kg = 20 kg. So, each side of the scales should be 20 kg ÷ 2 kg = 10 kg. It is possible if we put 5 kg + 3 kg + 2 kg on one side and 4 kg + 6 kg on the other, and it matches the picture. Without the 3 kg weight, the total weight will be 1 kg + 2 kg + 4 kg + 5 kg + 6 kg = 18 kg. So, each side should be 9 kg. This would be possible if we put 5 kg + 4 kg on one side and 6 kg + 2 kg + 1 kg on the other, but the picture shows the 5 kg weight on one side with two other weights and the 6 kg on the other with only one other weight, so this does not match with the picture.

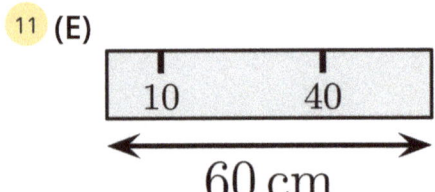

11 (E)

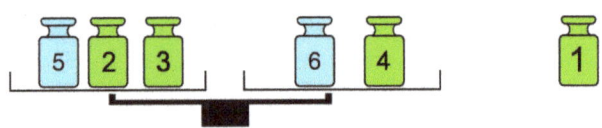

With (E) Andrew already has the length 10 cm marked. He can measure 20 cm by using the portion from 40 cm to the end, because 60 − 40 = 20. He can get 30 cm between the two marks because 40 − 10 = 30. The length 40 cm is already marked, and 50 cm is from the far end to the 10 cm mark as 60 − 10 = 50 cm. Finally, the length 60 cm is the whole ruler. Rulers (A) and (C) can't measure 40 cm. Rulers (B) and (D) can't measure 30 cm.

12 (A) 4

If we count the houses that are north of Road A and the houses that are south of Road A, this will be all the houses. So, there are 7 + 5 = 12 houses in total. Similarly, each of the houses is either east of Road B or west of Road B. Since there are 8 houses east of road B and there is a total of 12 houses, there are 12 − 8 = 4 houses west of Road B.

13 (D) 5

Since 19 is not divisible by 3 nor by 2, there is at least 1 car with two people and at least 1 car with three people.

If there was exactly 1 car with three people, there would be 8 − 1 = 7 cars with two people each. However, 1 × 3 + 7 × 2 = 3 + 14 = 17, so that would not be the right number of people.

If there were exactly 2 cars with three people each, there would be 8 − 2 = 6 cars with two people each. However, 2 × 3 + 6 × 2 = 6 + 12 = 18, so that also is not right as there are 19 people.

If there are 3 cars with three people each, there must be 8 − 3 = 5 cars with two people each. 3 × 3 + 5 × 2 = 9 + 10 = 19, so this arrangement works.

So, there are 5 cars which contain exactly 2 people each.

14 **(D) D**

Solution 1: After 10 stops, the train is at exactly the same position and is again traveling towards C (to the east). So, we only need to calculate the location after 96 − 90 = 6 stops. That brings it to D.

Solution 2: Start by marking the stop numbers. Note that the first stop is at C, and that the stops at each end only get counted once each time.

A	B	C	D	E	F
		1	2	3	4
9	8	7	6	5	(4)
(9)	10	11	12	13	14
19	18	17	16	15	(14)
(19)	20	21	22	23	24

We can continue this until the number 96 is reached, or we can notice that there are patterns. For example, the numbers for F are: 4, 14, 24, 34, 44, 54, 64, 74, 84, 94. Stop number 96 is going to be two stops to the left from F, so it will be D. Or, notice that for D in every other row there is 6, 16, 26, 36, and so on, so that 96 will also be under D.

15 **(B) 3**

Label the circles as shown in figure.

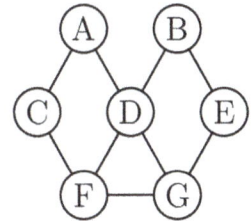

Circle D is connected to the most circles (4), so we have to start with this circle to get the correct answer. Let's start by painting it blue. Then we can use red for Circle F and green for Circle G, since those three need to be different colors. We can choose red or green for Circle A. If we choose red for Circle A, we can use green for Circle C because red is also used for Circle F and it is connected to Circle C, so we are still using only 3 colors. In the same way, we can choose red or green for Circle B. If we choose green, we can use red for Circle E because green is also used for Circle G and it is connected to Circle E. We now have colored all the circles using only 3 colors.

16 **(A)**

Sam has to walk along the numbers from the entrance to the exit.

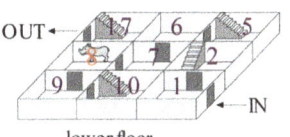 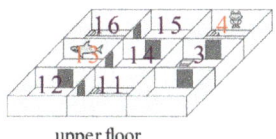

lower floor   upper floor

First, Sam will see the frog in room 4, then the rhino in room 8, and finally the shark in room 13.

5 Point Solutions

17 **(C) 3**

If number 1 is a kangaroo, then 2 and 3 are beavers. Then, if 2 and 3 are beavers, then 4 is a kangaroo. Similarly, 5 and 6 are beavers and 7 is a kangaroo. This is impossible because we only have two kangaroos, so 1, 4, and 7 can't be the kangaroos. Similarly, 2 cannot be a kangaroo because then the kangaroos would be 2, 5, and 8, and there would be three, and not two, kangaroos. Number 3 is a possibility. Then, 3 and 6 are kangaroos and the rest are beavers. Looking at the other answer choices, number 4 cannot be a kangaroo because then 1 and 7 also would be kangaroos, so there would be three kangaroos. Also, 5 can't be a kangaroo because then there would be three kangaroos (2, 5, and 8). So, only the numbers 3 and 6 are kangaroos.

Or: Among 1, 2, and 3, exactly one is a kangaroo. Among 6, 7, and 8, exactly one is a kangaroo. Among the 8 animals there are only two kangaroos, so 4 and 5 are beavers. Among 3, 4, and 5, exactly one is a kangaroo. So, 3 is a kangaroo. Likewise, among 4, 5, and 6, 6 is a kangaroo. Therefore, 3 and 6 are the only kangaroos. 6 is not an answer choice, but 3 is answer (C).

18 (B)

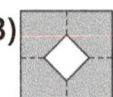 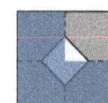

The upper right quarter of the square is the part that stays in place during folding, and it has a triangle cut out at its bottom left. So, the cut out part is in the center of the paper after unfolding. This is figure (B).

19 (B) 2

If Hermione is second, Ron must be first and Harry third. If Hermione is third, Harry must be first and Ron second. Therefore, there are two possible orders.

20 (D) D

First, let's note that each clock shows the time on the hour, and the times shown are 6:00, 2:00, 4:00, 3:00, and 7:00.
If one clock shows the correct time, one clock is ahead by an hour (is fast), and one clock is an hour behind (is slow), these three clocks will show three consecutive hours. The only three times that work are 2:00, 3:00, and 4:00. Since one time is an hour behind, one an hour ahead, and one shows the correct time, the middle of these times is the correct time. This is 3:00, and it means that clock D shows the correct time.

21 (B) 4

Brenda has 9 marbles. Since she has twice as many blue marbles as red marbles, she has 6 blue and 3 red marbles. She and Adam together have 10 blue marbles. This means Adam has the remaining 10 − 6 = 4 blue marbles.

22 (B) RSR

Note that the small black square on the sheet helps track how many times the paper is turned. It goes from the lower left to the upper right corner, so the paper had to be turned twice. Also, the stamp is made right side up, but is shown rotated once on the final paper. This means that exactly one of the turns was made after the paper was stamped. So, the machines were used in the order RSR.

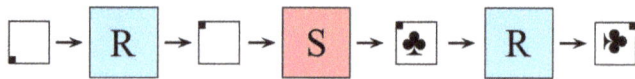

23 (D) 4

The problem can be solved by figuring out where the number 1 goes. Notice that number 1 cannot give a sum of 9 together with any of the available numbers because the biggest number is 7. The only circle that does not have 9 on either side is the circle at the very top, so the number 1 must go there. Then, we just need to fill in the numbers in the other circles going around in either direction until we get to the green circle. For example, if we go clockwise, we know that the next circle has the number 7 because 8 − 1 = 7. Then, we know that we need to write 2 in the next circle because 9 − 7 = 2. Finally, we see that the number in the green circle is 4 because 6 − 2 = 4.

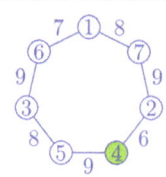

24 (E)

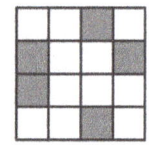

There are three shaded cells which are common for all grids. Looking at the differences, the only grid which differs from each of the other grids by exactly one cell is grid (E).

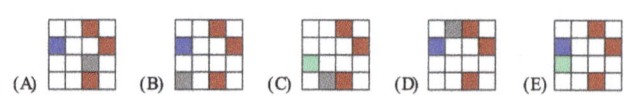

Part III
Answer Key

	1999	2001	2003	2005	2007	2009	2011
1	C	C	E	C	C	E	A
2	E	D	C	A	A	C	C
3	B	D	D	B	C	B	C
4	C	C	A	B	C	A	E
5	D	B	C	B	C	B	B
6	E	E	D	D	E	B	A
7	C	E	C	E	C	C	B
8	B	E	C	C	B	D	D
9	E	C	C	C	C	C	C
10	B	D	C	B	C	D	B
11	C	E	D	B	A	A	B
12	A	E	A	C	C	B	C
13	D	C	C	D	A	B	E
14	A	B	A	D	B	D	A
15	D	C	B	E	B	B	D
16	D	A	C	D	A	A	C
17	D	C	D	B	E	E	A
18	C	C	E	B	B	D	C
19	C	D	E	A	C	B	C
20	A	A	A	D	A	D	C
21	A	A	B	C	A	A	E
22	D	B	D	E	B	B	D
23	C	D	A	B	D	E	A
24	A	E	B	E	E	B	E

	2013	2015	2017	2019	2021	2023
1	D	E	E	E	C	D
2	D	A	A	C	C	C
3	E	E	E	A	B	B
4	B	C	C	D	B	C
5	C	E	A	A	E	C
6	D	E	E	A	C	E
7	B	B	C	D	D	C
8	E	D	A	B	C	E
9	B	A	E	B	A	A
10	D	D	E	C	E	A
11	D	B	B	C	E	E
12	A	B	C	E	C	A
13	D	C	D	B	C	D
14	E	B	D	A	E	D
15	B	D	C	D	D	B
16	B	C	D	D	B	A
17	D	A	C	E	D	C
18	D	C	B	C	E	B
19	B	A	E	B	D	B
20	C	D	D	E	A	D
21	E	E	C	B	D	B
22	B	C	B	E	E	B
23	B	D	E	B	D	D
24	B	C	B	D	A	E